Principles of

HAZARDOUS

MATERIALS

MANAGEMENT

by ROGER D. GRIFFIN

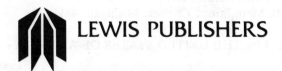

LEWIS PUBLISHERS

Library of Congress Cataloging in Publication Data

Griffin, Roger D.

Principles of hazardous materials management.

Bibliography: p.
Includes index.
1. Hazardous substances — United States — Management.
2. Hazardous substances — Health aspects — United States.
I. Title.
TD1040.G75   1988   604.7'0973   88-9391
ISBN 0-87371-145-9

Fourth Printing 1990

Third Printing 1989

Second Printing 1989

LEWIS PUBLISHERS, INC.
121 South Main Street, Chelsea, Michigan 48118

PRINTED IN THE UNITED STATES OF AMERICA

To my wife, Avice Marie,
and my children: Joel, Joshua, James, Benjamin, and Anastasia

Roger D. Griffin received his BS from California State University at Long Beach and his MS from the University of California, Irvine in 1969 and 1976, respectively.

Mr. Griffin is a registered Professional Engineer in Chemical Engineering in California. Additionally, he is a Diplomate in the American Academy of Environmental Engineers and a Certified Hazardous Materials Manager, at the Master level, with the Institute of Hazardous Materials Management.

From 1969 to 1977, Mr. Griffin worked with local government agencies, including the County of Orange (California), as well as the South Coast Air Quality Management District and its predecessor agencies. He served in various capacities as a lab chemist analyzing well water, soils, and air samples for trace pollutants, as a field inspector, as a source testing specialist, and as a permit processing engineer.

For the past 12 years, Mr. Griffin has worked as a consultant in the air quality management, solid waste, geotechnical, and hazardous materials fields. He has worked for private industry and government agencies in technology assessments for emissions control, regulatory impact analyses, permitting assistance, hazardous waste incineration, process evaluations, air quality dispersion modeling, solid waste-to-energy projects, chemical and low-level radioactive waste treatment and disposal, Superfund site remedial investigations and feasibility studies, and assessments of property contamination for real estate transactions.

Mr. Griffin has published articles and presented papers on low-level radioactive waste incineration, dioxin formation from municipal solid waste combustion, and air emissions from industrial operations.

He is a member of the Air and Waste Management Association and has served on the solid and biomedical waste committee. He is also a member of the UCLA Advisory Board for the certificate program in Hazardous Materials Management and has taught the core course in basic principles of hazardous materials management since the inception of the program.

PREFACE

This book was written to provide an introduction to the various disciplines that are concerned with hazardous materials. How chemical contaminants affect human health (Chapters 1 through 3), and how they are transported (Chapters 4 through 6), measured (Chapter 7), managed (Chapters 8 and 9), and regulated (Appendices) are the topics covered. They provide a framework for understanding the multidisciplinary nature of hazardous materials and how the risks to human health from such materials may be reduced.

This book was born out of the author's frustration in teaching an introductory course in the principles of hazardous materials management at UCLA in the mid-1980s. There were (and are) many detailed, discipline-specific texts available, but they were generally beyond the scope (or pocketbook) of the average person wanting to understand the basic principles in the field. Hence, this book was conceived and written.

It attempts to bridge the gap between popular articles and detailed scientific material. The audience is intended to be people with a general interest and those practitioners in hazardous materials management who are working to expand their understanding of the other disciplines. It is hoped that both groups will find that this book meets their needs.

ACKNOWLEDGMENTS

The following individuals are acknowledged for their contributions to this book and to my understanding of the field of hazardous materials management:

Michael Francis of Rockwell International

Dr. Kathryn Kelly and Dr. Joyce Tsuji of Environmental Toxicology, International

Jeff Thomas and George Skladany of Detox, Inc.

Wayne Tsuda of the County of Los Angeles

Paul Kaplow of Maler and Kaplow

Eugene Tseng of UCLA and American Ecology

Dr. Stu Salot of CTL Environmental Services

Jay Mackie, Alison Gemmell, and Dr. Tom Higgins of CH2M Hill

Bill Bingham and Meredith Meiling of BFI/Hospital Waste Systems

A special acknowledgment is given to my wife, Avice Marie Griffin, for her patience and understanding, without which this book would not have been possible.

CONTENTS

1 INTRODUCTION TO HAZARDOUS MATERIALS

Throughout history, people have been concerned with their health. These concerns broadly fall into three categories: the health of the general population, occupational hazards, and the availability of food or water supplies.

In ancient times, the general population was mostly concerned with preventing disease and physical injury. Ancient codes from 2000–1500 B.C. had laws regulating actions to be taken to deal with both disease and injury. The codes of Hammurabi and the laws of Moses were fairly detailed and specific in their instructions and are representative of the concerns of ancient societies.

Occupational hazards were known in ancient times to directly affect health, but they were limited in the number of persons affected. Lung diseases, tumors, blindness, cancers, and early death were among the typical occupational hazards. Mining and metalworking were of particular hazard, so slaves or conquered tribes were forced to perform those duties. Since mineral resources (copper, tin, lead) were considered vital and the average person was not generally exposed to the fumes and toxic dusts of such activities, these occupations continued.

However, food and water supplies in ancient times were of concern to all members of society. Perceived or actual threats to the food supply were sufficient to galvanize the efforts of an entire society to conserve what remained and make provisions for even harder times anticipated in the future. Joseph, the biblical prime minister in ancient Egypt, ordered one-fifth of all grain supplies to be appropriated and stored in huge granaries built on his orders in anticipation of famine.

1

The effects of contaminated land on agriculture were known in ancient times, since it was a common practice to sow an enemy's cities and fields with salt. Salt, of course, effectively ruined the food supply and usually necessitated relocation.

Water supplies were critical for both nomadic and agricultural societies. Indeed, many local wars were fought over water supplies. When water supplies became contaminated, the usual practice was to move to "greener pastures," or to seek other sources. In Exodus 7:24, for example, the people dug water wells when the River Nile temporarily became tainted and undrinkable. The search for pure water in arid lands was a major ongoing effort for early societies.

The situation remained this way for the better part of recorded history. With food and water supplies secured, and hazardous occupations consigned to slaves or subjugated tribes, health concerns for the average person remained concentrated on disease and physical injury before the coming of the Industrial Age. (The health effects of using lead pipes for water supplies or using red lead oxide in cosmetics were, of course, largely unknown.)

In the early 19th century, the health problems normally associated with occupational hazards began to affect the common person. By the early 20th century, exposures to minerals and dusts from smelters, gases from coal and oil combustion, and vapors from oil refining and chemical processing caused the average citizen to begin feeling the effects formerly experienced occupationally by industrial workers. In an ever-widening circle, health impacts among the general population began to move from acute effects (injury, short-term illness, death) to chronic or long-term effects (cancer, paralysis, "consumption").

Attempts were made to reduce the more obvious public hazards by legislation abating smoke emissions or limiting dumping of wastes into rivers or harbors. However, occupational hazards and threats to food and groundwater supplies largely had to wait another 50 years.

By the mid-20th century, attempts to address acute health risks were generally successful and the emphasis began to shift to chronic hazards. The use of asbestos as a fireproofing agent in construction is a case in point. Though successful in reducing deaths from fires, asbestos has been implicated as a cause of lung cancer, both to workers installing it, and to the general public occupying buildings where these materials were used.

Another example is seen in the clothing dry cleaning field, where flammable solvents have been replaced with nonflammable solvents such as perchloroethylene. Death and injury from fires and explosions in dry cleaners have decreased, but with the passage of time increased concerns have been raised for chronic health effects such as headaches, dermatitis, and cancer. These are direct concerns for workers using these solvents and they have become indirect concerns to the rest of society, since these nonflammable solvents have been found as contaminants in drinking water supplies.

Other attempts to enhance food supplies by using additives (diethylstilbestrol, known as DES) to fatten beef have led to concerns for long-term health effects such as birth defects.

These occurrences have led to a more holistic approach to health in society today — one in which materials that may present a hazard are to be evaluated throughout their life cycle. Today the challenge is to determine the risks of potentially hazardous materials from the point of origin through usage to final destination, whether in a discarded material or as a trace contaminant in air, water, or food supplies.

The principles of hazardous materials management presented in this book provide a means of assessing those risks through a study of the toxicology of those materials and a risk assessment process. The means by which hazardous materials may be disseminated into the environment (air, groundwater, transportation) and the management approaches that may be utilized in dealing with those risks (analysis, minimization, waste treatment, disposal) are also examined. By understanding these areas, it will be possible to reduce the risks associated with hazardous materials.

2 PRINCIPLES OF TOXICOLOGY

"All things are poisons, for there is nothing without poisonous qualities. It is only the dose that makes a thing poison." With this statement over 400 years ago, Paracelsus expressed a major fact of toxicology. It is still true today when one is considering the health impact of hazardous materials. In environmental toxicology, one is concerned with poisons that occur in the ambient environment to which human beings may be exposed. This includes exposures to toxic agents in the workplace and food chain as well as natural and manmade activities.

The basic principle of toxicology is that the size of the dose determines the effect. Dose refers to the ratio of the mass of toxic substance administered to the body weight of the individual and the time over which that dose is administered. Coffee can be lethal (due to the effect of caffeine, a very powerful toxicant) if enough cups are drunk in a short time period.

Lifestyle can have as much impact on toxic effect as environmental factors. The difficulty in the field of environmental toxicology is distinguishing between exposure to a particular toxic material through nonvoluntary pathways such as air pollutants or contaminated fish and exposure brought about by lifestyle (i.e., use of alcohol, tobacco, and drugs).

This chapter will look at the basic principles of toxicology: dose-response testing, major routes of exposure, body response to various toxic chemicals, varieties of toxic effects, and allowable exposure concentrations that have been published as standards by various governmental agencies.

DOSE-RESPONSE

Toxicity tests are at the heart of environmental toxicology. Since it is generally illegal to expose people to suspected toxicants, other species are used to quantify the effects that may be expected from exposure to hazardous or toxic substances.

In these tests, a group of test animals is exposed to a measured substance under controlled circumstances and the response (either death or other end points) is monitored. The tests are repeated at different doses with new test animals over a period of time and the results are observed and logged. Simultaneously, a control group of roughly the same number, sex, and species of test animals is also monitored. The end result of such a test is a dose-response curve.

A generalized dose-response curve is seen in Figure 2.1,[1] which could be the response of rats to a certain chemical. From this curve one may determine a response, such as the LD_{50} for rats to that chemical. (LD_{50} is the lethal dose for 50% of the test animals. Other end points are possible, depending on the tests run.) Of particular importance is the shape of the curve and the implications thereof. First, it is an S-shaped curve, which indicates that effects are different for incremental changes in dose. Second, it does *not* go through the origin (zero response at zero dose). This

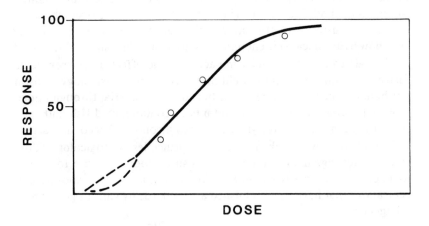

Figure 2.1.   Generalized dose-response curve showing low dose extrapolations.

indicates that the true effect at very low doses is unknown (or maybe zero response for some minimal dose). These important points are examined in detail below.

Several things should be noted about generating dose-response curves.[2] First, higher doses are normally chosen in order to speed up the response time. Second, it may cost up to $100,000 to get one test point for one species. Therefore, only a few tests are run at rather high doses; then the curve is extrapolated down to low doses. Extrapolation, at best, is an art form, so how one draws the lower portion of that curve is left to the investigator. The significance of this is explored later, but it is important since environmental exposures are generally much smaller than doses found in either a laboratory or a workplace.

For the lower dashed line in Figure 2.1, one may see essentially no response until some moderate level of dose is reached, at which the effect becomes truly observable. The upper dashed line is a linear extrapolation between the lower test points and the zero point on the curve. This extrapolation presumes that there will always be a response for a given dose. This presumption has been verified in the exposure of lung tissue to asbestos fibers. Other extrapolations are usually based on assumptions on the part of the investigator.

Major species differences do exist, and this presents the biggest problem in trying to extrapolate dose-response, not only at lower dosages, but also between species. Absorption rates, metabolic activity, and excretion rates are all included in the differences between species. Individual differences in the same species can have a dramatic effect on the response of the test animal. Habitat (e.g., aquatic vs terrestrial) is also important. Such features as genetic traits, sex and hormones, nutrition, and age of the test animals in some degree give a scatter in the response data to a given dose. Rats are typically used because they are cheap, have a relatively short life span, and are mammals.

Figure 2.2 shows a typical series of potential dose-response curves that one might find for different chemical toxicants. This illustrates some of the problems with dose-response curves at low doses. At small doses, one is left with the question, "Which is the more toxic of the three different chemicals seen: A, B, or C?" For a given (50%) response, chemical A is more toxic than B since the 50% response shows up at a much lower dose. However, for very small doses (say a 10% response) C would be more toxic than B, since the curves cross over. In both cases, A is the most toxic.

The four most important parameters in comparing dose-response curves and attempting to draw conclusions are species, dose, time period,

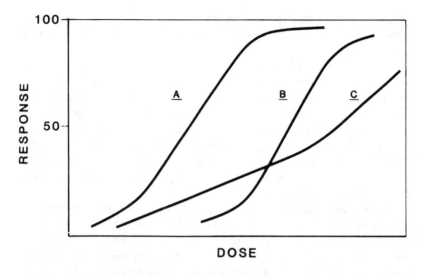

Figure 2.2.   Potential dose-response curves for different chemical toxins.

and end point chosen. The biggest problems in interpretation are related to interspecies differences in response to any given chemical.

A common example of such a difference is aspirin. It is a teratogen for rabbits (i.e., it causes birth defects in rabbit offspring), but it is generally recognized as safe for people. Arsenic is carcinogenic for humans but not for animals (even though it is toxic at certain doses to all). With respect to the chemical 2,3,7,8 TCDD (dioxin), there is a 5000-fold difference between its toxicity to hamsters and its toxicity to guinea pigs.

ROUTES OF EXPOSURE

The four major routes of exposure to hazardous materials are ingestion, inhalation, dermal absorption, and injection. Within the environmental toxicology field, the latter is a minor route of exposure to environmental contaminants as compared to the other three. Since injection is most often used with experimental animals, the results of experiments using injection may not be comparable to the effects from other routes of exposure.

Ingestion

Ingestion is the route of entry which deals with eating. Foods or other swallowed material (whether on food or mixed with fluids or contaminants occurring in food or drinks) are carried directly into the digestive tract and are metabolized. Contaminants may be carried from there throughout the body to target areas where the effects may be felt. The most striking example of an environmental toxicant using this route would be the methyl mercury poisoning at Minimata Bay in Japan. In this episode, dozens of fishermen and their families either died or suffered severe central nervous system damage from the effects of methyl mercury in their diet.[3]

The solubility of contaminants as they occur in food is a factor to be considered. Fat-soluble or lipophilic chemicals, such as benzene or DDT, are quickly absorbed into the body through the ingestion route.[4] Water-soluble (hydrophilic) chemicals may be absorbed throughout the body since the human metabolism operates on a water-based chemistry.

Chemical interaction with gastric fluids during digestion may alter the chemicals' solubility, and therefore their transport and metabolism in the human body.

Inhalation

Inhalation, the major route of entry when dealing with hazardous materials, involves absorption of airborne chemicals during breathing. The solubility of the chemical of interest in the blood affects the degree of its absorption. Once in the blood via the lungs, it goes directly to the brain and the rest of the body.

The most common example of environmental exposure by this route is the absorption of carbon monoxide from industrial processes or through smoking. Carbon monoxide has several hundred times the binding affinity for hemoglobin as does oxygen. Hemoglobin is the iron-based organic compound that is responsible for all oxygen transport in the blood. Oxygen is "bound" by hemoglobin and is later given up by hemoglobin to the tissues. When oxygen is displaced irreversibly by carbon monoxide, the transport phenomenon is not able to function and oxygen starvation of cells begins.

The retention of airborne particulates that carry or may carry toxic chemicals is highly dependent upon particle size due to the structure of the human lung. The smaller sizes penetrate deeper and have a greater effect.

Dermal Absorption

Dermal or skin absorption is the third major route of environmental exposure. Those areas of the skin that are in contact with chemical contaminants or materials transporting chemical contaminants are the most impacted. The degree to which skin absorption causes a health effect is roughly proportional to the fat or oil solubility of the chemical. Another major factor in the importance of the dermal exposure route is that it is specific to certain regions of the body. The face or hands are the areas of the skin surface that usually come into contact with potentially toxic materials.

Dermal absorption may also be enhanced by scratched, broken, roughened, or abraded surfaces of the skin on the ankles, hands, neck, or facial areas.

BODY RESPONSE TO TOXIC CHEMICALS

Chemical toxicants or carcinogens have a different effect on different member organs of the body depending on the dose and route of exposure. Other chemicals target specific organs depending upon oil or fat solubility, the effect that the chemical may have on enzyme activity, or physical interruption of the transmission of electrical impulses.

Table 2.1 lists a variety of systemic poisons[5] that influence different target organs in the human body by type of chemical or hazardous substance. The hepatotoxic agents, such as carbon tetrachloride, affect the liver primarily, whereas the halogenated hydrocarbons affect the kidneys. The neurotoxic agents that affect the nerve system include methyl alcohol, carbon monoxide, heavy metals, and organometallic compounds. The hematopoietic toxins affect the blood or blood cells and consist of aromatic compounds such as benzene, phenols, aniline, and toluidine. The anesthetic or narcotic chemicals (which affect consciousness) consist of ketones, aliphatic alcohols, and double-bonded or "ether" types of organic compounds.

Table 2.1 Systemic Poisons and Target Organs

Hepatotoxic agents (liver)

Carbon tetrachloride
Tetrachloroethane

Nephrotoxic agents (kidneys)

Halogenated hydrocarbons

Hematopoietic toxins (blood)

Aniline
Toluidine
Nitrobenzene
Benzene
Phenols

Neurotoxic agents (nerve system)

Methanol
Carbon disulfide
Metals
Organometallics

Anesthetics/narcotics (consciousness)

Acetylene hydrocarbons
Olefins
Ethyl ether, isopropyl ether
Paraffin hydrocarbons
Aliphatic ketones
Aliphatic alcohols
Esters

Source: Clayton, G. D., and F. E. Clayton, *Patty's Industrial Hygiene and Toxicology*, vols. 1 and 2 (New York: John Wiley and Sons, Inc., 1986.)

The Lungs

The lungs are the most quickly impacted of all of the body organs to toxic contaminants. This is due to the fact that the lungs are in constant contact with the environment. Also, the lungs may have a surface area of 70-100 m², as opposed to the skin at 2 m², or the intestinal tract at about

10 m². This becomes an important point since the rates of absorption of various contaminants are considered to be a direct function of surface area exposed to a contaminant.

Figure 2.3 presents the various passages and sections of the human lung. The lung acts as a particle filter by its varied construction. As one goes deeper in the lungs, one finds more narrow and more tortuous paths which present an aerodynamic obstruction to particle movement. Particle size, therefore, determines where particles will be deposited in various portions of the lung (the smaller, the deeper). Gases, of course, penetrate to the deepest segments of the lung.

The nose and pharynx will capture particles from 5 μm to greater than 50 μm in size. The larynx down to the bronchi regions of the lungs will collect particles from 1 to 5 μm in size. The smaller portions of the lungs, the bronchioles and alveoli, will collect particles in the 0.5 μm range. (Particles smaller than 0.05 μm will be exhaled.)

The filtering construction of the lung is helped by the fact that mucus covers the upper reaches of the lung passages. Particles that are deposited in the branched passages of the upper reaches of the lung will be removed by a phenomenon called the "mucociliary elevator." This is a movement of mucus upward and out of the lungs caused by small ciliary hairs in the lung passages. Thus, inhaled material will be moved up out of the lungs by this "elevator" effect after the material has been captured

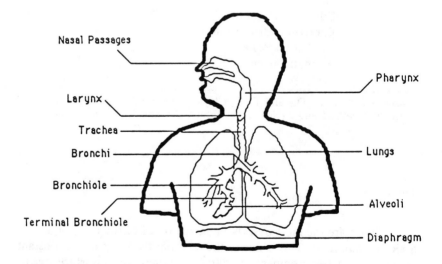

Figure 2.3.   A cross section of the human respiratory system.

in the mucus. (Note that inhaled material could potentially reach the stomach from the lungs due to a swallowing action.)

Down in the alveoli portions of the lungs, only the small particles can directly enter. But once in, they do not depart — the alveoli are not coated with mucus. Damage may thus occur in the lung itself. Inflammation and irritation to the lungs is also possible, which may cause the lungs to be scarred and toughened. The alveoli or tiny air sacs may lose their elasticity due to repeated impact by particles and the scarring action of gases, paints, or solvents. Emphysema is one result.

Certain contaminants or even small dissolvable particulates may enter the blood stream directly and then be transported to target organs such as the liver or kidneys for their effect.

Lung cancer is the one environmental effect which has been established as being directly produced by an asbestos particle in the lung. This case has been documented so that some models of cancer-causing mechanisms have a basis in reality (the one-hit model).

The Skin

The skin is the largest organ by weight in the human body and it also is in relatively constant contact with the environment. The skin serves to keep toxic agents, chemicals, and microorganisms outside of the body and also protects from loss of water by the cells on the inside of the body. This action is accomplished by the skin presenting to the "outside world" an oil-based layer that does not allow water-based environmental toxicants to penetrate. This has been effective throughout history, since pre-Industrial Age toxic agents were generally water-based.

With the advent of modern technology, oil- or solvent-based chemicals have come into wide usage. These chemicals are readily absorbed into the skin since, being fat- or oil-soluble contaminants, they will dissolve directly into the lipophilic layers of the skin. A common example is when one spills gasoline on the hands. Even after repeated washing with soap and water, the odor of gasoline can still be detected for a long period of time. This odor is due to "out-gassing" of gasoline that has dissolved into the subsurface areas of the skin.

Figure 2.4 illustrates the various layers of skin, the defense mechanisms of the skin, and the injuries that may be suffered by the skin by various means. These include chemical attack, disease, and physical trauma.

The skin, as an organ, maintains an equilibrium between its internal

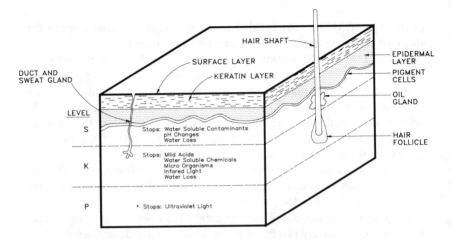

Figure 2.4.   An enlarged cross section of human skin.

structures and the external environment by sweat glands (heat equilibrium by evaporation), by pigment cells (protection from UV light), and oil glands that excrete oil to cover and soak into the surface of the skin.

The skin is not a perfect barrier, and absorption of toxic agents will be enhanced by breaks in the skin, increased concentrations of environmental contaminants, and decreased particle sizes. The skin serves to localize effects, since there is no transport medium from one portion of skin to the other. However, systemic effects may be observed once a contaminant has passed through the skin into the blood system.

The Eyes

The eyes are in much the same position as the skin; however, they serve a much more critical function, i.e., vision. They are much more sensitive than the skin to chemical reaction and are prone to damage in accidents such as acid or alkali splashes. Physical media such as particles (which can abrade or penetrate the eyes' surfaces) and ultraviolet and ionizing radiation have serious effects on the eyes.

The Central Nervous System

The central nervous system (the brain, spinal column, and nerves) is generally impacted by dissolved heavy metals such as mercury and lead.

The Minimata disease described earlier is a prime example of central nervous system damage caused by exposure to a contaminant, as is the "Mad Hatter's disease" of the 19th century. The Mad Hatter in *Alice in Wonderland* suffered, as did most hatters, from mercury poisoning caused by a high level of mercury in the tanning solutions used to make hats for men and women.

The Liver

The liver is a very serious target organ for cancer since it is a metabolic center for the entire body. For example, it is in the liver that carbon tetrachloride is converted to chloroform, which is a carcinogen and toxic to the cells with which it comes in contact. Sufferers from other liver dysfunctional diseases, such as alcoholism, are acutely impacted.

The Kidneys

The kidneys are the main filtering media in the body and thus have a very high exposure to toxicants. As a filter, they serve to concentrate certain contaminants, such as heavy metals and halogenated hydrocarbons. As the concentration increases, the "dose" becomes greater and toxic or carcinogenic responses increase.

The Blood

The blood system is impacted by agents such as carbon monoxide that affect the oxygen-carrying capacity of hemoglobin. Direct blood cell impacts also occur due to chemicals such as the aromatic compounds benzene and toluene, and phenolic compounds.

The Reproductive System

The reproductive system is prone to impact by environmental contaminants since it is a center of DNA activity. Thus, any contaminants that affect cell division and the transmission of genes will affect the reproductive system. There are significant differences between the sexes regarding their response to reproductive toxins. Fertility, sperm count, and cancers are all affected by environmental contaminants. Lead and DES have decided effects on the reproductive system.

Other Factors to Consider

Table 2.2 gives a relative index of toxicity that may be used when considering the oral lethal dose for human beings. While this is not a specific reference table, one may get some idea of the amount of a toxicant it would take to be lethal to an average human being. Elderly, young, and smaller-bodied individuals would show an enhanced effect over the average.

Individual differences may thus have a profound effect on response to toxic chemicals. Genetic makeup controls the presence or absence of key enzymes that are the biochemical catalysts for metabolism. Humans are highly heterogeneous and individual differences are widespread.

Sex and hormones make for different responses to toxicants. Hormones influence enzyme levels, and therefore chemical toxicity. Thus, males and females are often not equally sensitive to chemical doses, and pregnant females are in some instances affected differently than non-pregnant females. Differences in physiology between males and females can also influence response. For example, females have a higher body fat percentage, and could be more adversely affected by fat-soluble contaminants than males.

The body's adaptive mechanisms to nutritional and dietary stress may result in increased sensitivity to, or uptake of, some toxic chemicals. For example, a low-mineral diet increases the uptake of iron, calcium, and zinc, but also increases the uptake of toxic metals such as cadmium and barium.

Table 2.2 Relative Index of Toxicity

Toxicity Rating or Class	Probable Oral Lethal Dose for Humans	
	Dose	For Average Adult
1. Practically nontoxic	>15 g/kg	More than 1 quart
2. Slightly toxic	5–15 g/kg	Between 1 pint and 1 quart
3. Moderately toxic	0.5–5 g/kg	Between 1 ounce and 1 pint
4. Very toxic	50–500 mg/kg	Between 1 teaspoonful and 1 ounce
5. Extremely toxic	5–50 mg/kg	Between 7 drops and 1 teaspoonful
6. Super toxic	<5 mg/kg	A taste (less than 7 drops)

Infants and children are often more sensitive due to undeveloped tissues, and possess a reduced ability to metabolize and detoxify chemicals. Additionally, they may have a much lower body mass, so for a given intake the dose would be much higher than for an adult. The elderly may be more sensitive to toxic chemicals due to the reduced detoxifying capacity of the liver and excretory capacity of the kidneys. Susceptibility to injury or other aging factors may also increase sensitivity to environmental toxicants.

The different routes of exposure to a chemical carcinogen or toxicant will have a decided effect on how the body reacts. For instance, nickel fumes are carcinogenic if inhaled through the nose or larynx, but if nickel fumes are ingested, they are not carcinogenic.

TYPES OF TOXIC EFFECTS

There are a wide variety of bodily responses to environmental toxins that are of concern in the field of hazardous materials management. The following are the six types of toxic agents, examples of such agents, and the bodily responses they provoke:

- allergic agents — isocyanates: cause itching, sneezing, or rashes
- asphyxiants — hydrogen cyanide and carbon monoxide: displace oxygen
- irritants — hydrochloric acid, ammonia, and chlorine: may cause pulmonary edema at very high concentrations, or unpleasant sensations throughout the body
- necrotic agents — ozone and nitrogen dioxide: directly cause cell death
- carcinogens — asbestos, arsenic, and cigarette smoke: cause cancer
- systemic poisons — benzene and arsenic compounds: attack the entire body

Synergism and antagonism are two other concepts important to the understanding of toxic effects in the human body. Synergism is best summarized by saying two plus two equals five. This describes the enhancing effect of two different environmental contaminants upon the human body. For instance, asbestos may cause cancer, but the carcinogenic effect of asbestos is 240 times greater with a person who also smokes. Likewise, the narcotic effect of alcohol is greatly enhanced by the presence of carbon tetrachloride.

Antagonism also occurs, and is best summarized by saying two plus two equals three. An antagonistic effect exists where one impact is lessened by having another substance present. One may find that eating bread before drinking alcohol modifies the immediate effects of the alcohol compared to not eating bread with, or prior to, alcohol ingestion.

SPECIFIC CONTAMINANT EFFECTS

Carcinogens

Carcinogens are those contaminants which may cause cancer. On the average, cancers are considered approximately 15% related to genetic effects and 85% related to environmental effects (including both lifestyle and environment).

Mutagens

Mutagens are agents that cause genetic damage in reproductive cells. They act by alteration of the DNA structure of the genes and the effects are felt by the next generation. Radiation exposure is probably the most powerful example of a mutagenic effect.

Teratogens

Teratogens are agents that adversely affect offspring while in the fetal stage of development. Thus, offspring would be most likely to experience a teratogenic effect due to exposure of the mother to an environmental hazardous substance. A well-documented case of such a teratogenic effect was the birth defects found in children of women who ingested thalidomide-containing pills in the early 1960s. This effect was noted only when the pregnant females took thalidomide on the 12th day following fertilization. Thus, it had a developmental effect on offspring during the fetal period. Other teratogenic substances include alcohol, carbon monoxide, anesthetic gases, and DES.

Latency

A concept important to toxic effect is the latency period. This is the time between exposure of an individual and the clinical manifestation of an adverse effect. Mesothelioma (a chest cancer) is a prime example of latency. Workers involved in shipbuilding during World War II were exposed to airborne asbestos fibers at very high concentrations, yet the effect (cancer) did not show up until 20 to 30 years after the exposure had ceased.

EXPOSURE LIMITS

In the field of environmental toxicology it is recognized that some substances exhibit a threshold level. This is a dose below which no toxic or adverse effects may be seen. Other substances are assumed (where data are lacking) to have a zero threshold level of exposure; that is, any exposure above zero is considered to be negative or adverse. As related later in the section on analytical methods, we know that an absolute zero level of exposure does not exist. Nevertheless, federal and state regulatory policy assumes a zero threshold for carcinogens as the only safe dose.

With respect to a wide variety of noncarcinogenic contaminants, however, certain exposure levels are allowed by law.

The most well-known of these allowable exposure standards are the Threshold Limit Values (TLV®) published by the American Conference of Governmental Industrial Hygienists.[6] These are levels of airborne exposure concentrations at which adult healthy males working at 40 hours per week generally do not suffer adverse health effects. Unless adopted by a governmental agency, these are not government standards, but are recommendations made by industrial hygienists. The list is updated annually.

The TLVs are published in two levels: the time weighted average (TWA), which is the allowed eight-hour average concentration of exposure, and the short-term exposure level (STEL), the maximum 15-minute average concentration to which a worker may be exposed. In spite of disclaimers, both of these are finding their way into the hazardous materials field as standards. One may notice that the concept of "dose" is implicit by virtue of concentration and exposure time being specified for both the TWA and the STEL. The Occupational Safety and Health Administration (OSHA) has adopted certain of these TLVs as permissible exposure levels (PELs). PELs therefore become a legally enforceable level of exposure for industrial workers.

For lack of data, other agencies may adopt certain of these TLVs combined with a factor (less than one) to account for perceived differences in individual ability to avoid adverse health effects. In the air toxics field, it is not uncommon to see a factor of 1/100 or 1/1000 of the TLV being adopted as a permissible toxic air exposure concentration.

The "no observed adverse effect level" (NOAEL) is frequently used as the starting point of allowable exposure for potential human levels based upon animal studies. If such a level was 1000 ppm (inhalation) for rats for 12 months, an agency would typically divide the NOAEL by 10 to account for interspecies differences and divide that by 10 again for individual differences. This level is then commonly divided by another factor (from 10 to 100) due to uncertainty. This new level is assumed acceptable for human exposure due to the unknowns inherent in toxicological investigations. So, for this example, a 1000-ppm rat NOAEL level could become an "acceptable" level for a human of 0.1 to 1.0 ppm over a 12-month period.

Other "acceptable level" approaches use the "no observed effect level" (NOEL), the "acceptable daily intake" (ADI), percentages of existing background exposures, or, most recently, whatever dose may correspond to a given risk of cancer (one in 100,000 or one in a million). The latter is the topic of the next chapter.

REFERENCES

1. Kamrin, M. A. *Toxicology: A Primer on Toxicology Principles and Applications* (Chelsea, MI: Lewis Publishers, Inc., 1988).
2. Kelly, K., Environmental Toxicology International, Inc. Personal communication (Nov. 16, 1986).
3. "Occupational Diseases—A Guide to Their Recognition," U.S. DHEW, Center for Disease Control, NIOSH Pub. No. 77–181 (1977), p. 373.
4. Sittig, M., Ed. *Handbook of Toxic and Hazardous Chemicals* (Park Ridge, NJ: Noyes Data Corporation, 1981).
5. Sax, N. I., Ed. *Dangerous Properties of Industrial Materials,* 5th ed. (New York: Van Nostrand Reinhold Company, 1984).
6. "TLVs—Threshold Limit Values for Chemical Substances and Physical Agents in the Work Environment (Cincinnati, OH: American Conference of Governmental Industrial Hygienists, updated annually).

3   RISK ASSESSMENTS

INTRODUCTION

"Use of this product may be hazardous to your health. This product contains saccharin which has been determined to cause cancer in laboratory animals." This quotation, occurring on the label of every product containing saccharin, indicates that there is risk to a person who consumes that product. Yet the production and use of substances such as saccharin continue to this day. This leads to the question, "What is the risk to my health from this substance?" This chapter will look at what risk is and how it is assessed, especially risk to health.

Risk is sometimes defined as the expected value of an undesirable consequence or hazard. Table 3.1 indicates several consequences or hazards that may be faced in life with a measure of the risks associated with each. Death or cancer can be rated on an occurrences-per-year basis, whereas economic loss can be measured in terms of dollars lost per year. In the case of contaminated land, the acreage lost per year due to nonutilization is another measure. For these consequences and their measurements, a number of different fields may be impacted. Dollar losses impact the insurance field, deaths impact the automobile and transporta-

Table 3.1 Hazards and Measures of Risk

Consequence or Hazard	Measure of Risk
Death (from acute hazard)	Early deaths/year
Economic loss	$ Lost/year
Cancer death	Early deaths/year
Contaminated land	Acres cost/year

21

tion industries, and industrial accidents impact the industrial process industries. One must be careful, however, to distinguish between a measure of risk, such as the number of accidents or cancers per year, and its perception, which is a social-political activity (e.g., "chemicals are bad").

The perception of risk is a key element in understanding the utilization of risk assessments in decisionmaking processes. Risks are a part of life. Therefore, no amount of effort will eliminate all risk. Whether a person or a group accepts or rejects a potential risk is a decision that goes beyond risk assessment and becomes risk management. For example, aflatoxin, one of the most powerful toxic substances known, is produced naturally by molds growing on peanuts. Some studies indicate that many people would accept a potentially fatal risk of 1 in 10,000 by using peanut butter (which might contain aflatoxin), yet would reject utilizing a material that contained a chemical which might produce a cancer risk estimate of 1 in a million (10^{-6}) or 1 in 10 million (10^{-7}).[1] An assessment of the perception of risk, however, goes beyond the scope of this chapter.

When a risk assessment is performed, it is an attempt to derive a quantitative value of either a societal or individual risk. As noted above, there is a wide variety of risk assessments that may be performed in different industries including medical care, municipal utilities, and transportation.

Types of Risks

There are two general types of risks: a societal risk and an individual risk. A societal risk is the number of incidences or consequences that can occur per year. An individual risk is the probability of any single consequence occurring to an individual during a year.

The distinction is illustrated by data on automobile accidents and automobile deaths in the United States.[2] In 1985, about 15 million automobile accidents occurred, of which 1 in 300 resulted in the death of a person. The societal risk is a measure of the number of deaths that occur per year, or $1/300 \times 15$ million $= 50,000$ deaths/year. The individual risk is the probability of such an event or hazard occurring to any one person.[3] Given that the overall death rate is 50,000 per year and that there are 250 million people in the United States, the individual risk is approximately 50,000 deaths/year/250 million, or 2.0×10^{-4}.

Assessment vs Management

Risk assessments are actually part of a two-phased approach to dealing with a perceived hazard or consequence of certain activities. The first phase is the assessment of risk, either by actual numbers or probability. The second aspect is risk management, that is, a decision that can be made or action that can be taken to mitigate the risk or minimize the overall consequences of any particular activity.

In the environmental field, risk assessments are performed generally by practitioners in the disciplines related to the project or activity being considered (typically engineering or science). The management of that risk is usually a decision by a government agency as to whether a proposed project or activity will proceed. A more thorough approach would be an iterative process between the assessment and the management in which the ultimate aim is to minimize the risk of a particular activity.

Environmental Risk Assessments

For environmental risk assessments, one is typically concerned with the potential for a particular activity to cause chronic health problems in humans. Principally, these assessments deal with cancer rather than acute incidents such as injury or death. Unfortunately, for a number of chemical contaminants one is faced with a relatively small body of health data. Therefore, many assumptions and uncertainties are present.

Certain contaminants in some media are not subject to a risk assessment. These include criteria air pollutants that have established ambient air quality standards (AAQS). When these standards were set, appropriate safety factors were included in setting such standards. Drinking water quality standards were specifically set based on early risk assessments.

The approach to assessing risk dates back to Atomic Energy Commission (AEC) regulation of the nuclear power industry in the 1950s. The concern was to minimize the number of potential nuclear accidents that could occur at a facility based upon data for certain types of hardware and design systems. Such an approach deals with acute hazards due to failures of equipment. Nevertheless, the initial AEC approach to risk assessment formed the basis of modern techniques. Since that time, regulatory agencies have adopted formal guidelines for risk assessments for environmental chemical hazards and health impacts.

THE RISK ASSESSMENT PROCESS

Risk assessments can be performed for a variety of environmental contaminants. These include chemical contaminants, but may also include ultraviolet radiation, electromagnetic fields, and microwave radiation. In 1986, the Environmental Protection Agency (EPA) published formal guidelines to be followed for five areas of risk assessment.[4] These include the guidelines for carcinogens, mutagens, chemical mixtures, and suspected developmental toxicants. One guideline exists for the estimation of exposure.

In these guidelines and backup documentation are EPA policy and inferences with which one should be familiar when performing risk assessments.

Inferences are made based upon logic and reason, given what is currently known about metabolic activity in animals and humans. Models of carcinogenicity are set up based upon assumptions or inferences of biochemical activity.

As policy, the EPA holds that there is no safe threshold concentration for a chemical carcinogen.[5] Asbestos, however, is the only known carcinogen in which, potentially, one fiber may cause cancer.

The four major elements in a health risk assessment process, per the EPA guidelines, are:

1. hazard identification
2. dose-response assessment
3. exposure assessment
4. risk characterization

Figure 3.1 illustrates the key features in the risk assessment process.

Hazard Identification

The hazard identification part of health risk assessment is a qualitative assessment based upon a review of relevant biological and chemical information. Since chemical agents seldom occur in a pure state and are often transformed in the body, assessments should include information on contaminants, degradation products, and metabolites.

The elements involved in making a hazard identification of a given chemical are:

- physical and chemical properties of the agent
- routes of exposure
- structural or activity relationships that may support or argue against prediction of carcinogenicity
- metabolic properties of the agent
- toxicological effects of the agent
- information on both short- and long-term animal studies that have been performed to date

Since animal studies are frequently used in determining carcinogenicity in humans, these receive a great deal of weight in the hazard identification process. The weight of evidence that an agent is potentially carcinogenic for humans increases with:

1. the number of tissue sites affected by an agent
2. the number of animal species, strains, or sexes that will show a carcinogenic response
3. the occurrence of clear-cut dose-response relationships in treated (as compared to control) groups
4. a dose-related shortening of the latency period with respect to the agent
5. a dose-related increase in the proportion of tumors that become malignant or true cancers

The conditions of the experiment have to be reviewed carefully as they relate to the relevance of the evidence to human carcinogenic risks. Interpretation of animal studies data is very important as are latency period and dose-related changes that are noted. Evidence of an agent promoting or showing cocarcinogenic activity is sufficient to assume that the agent is a complete carcinogen in its own right.

There is a weighting scale applied to the evidence that is presented. Data are classified alphabetically by groups (A through E) in descending order of overall weight of evidence:

Group A—data that relates to known carcinogenic effect to humans

Group B—probably carcinogenic to humans

Group C—possibly carcinogenic to humans

Group D—not classified as to human carcinogenicity

Group E—evidence of noncarcinogenicity for humans

To evaluate carcinogenicity, the primary comparison is tumor response in dosed animals as compared to other controls. Specific organs or tissues are also to be included in this identification. Where epidemiological studies (on humans) are available, these results must be included in the data base for a particular agent. Where the data for human studies are strong and clear-cut, they become of paramount importance even over long-term animal studies (since we are concerned primarily with carcinogens in human beings).

Dose-Response Assessment

As noted earlier, the dose-response assessment is a critical element in determining the effect of a particular agent on carcinogenicity. The degree to which the agent is potent is important, as is its level of potency for a given dose, since these may vary widely among animal species and between animal species and humans.

Specifically, the data selected for a specific chemical should be organized as follows: first, human studies, and second, data from species that respond most like humans for different tissues and organs. Long-term animal studies showing the greatest sensitivity should generally be given the greatest emphasis.

However, when the exposure route from which the information is obtained differs from the route occurring in an environmental exposure, some consideration should be given when extrapolating from one route to another. Also, extrapolations to very low doses must be done carefully and in accordance with policy guidelines on modeling.

Low Dose Extrapolations

The choice of the mathematical extrapolation model to very low doses must be made carefully. Risks at very low exposure levels cannot be measured directly by either animal experiments or by epidemiological studies. Therefore, mathematical models are used to try to fit the observed data. Extrapolations may be made to doses lower than presented in animal studies.

No single mathematical procedure is recognized as the most appropriate for low dose extrapolation in carcinogenicity. When data and information are limited and when much uncertainty exists regarding the mechanism of carcinogenic activity, models or procedures that incorporate low dose linearity are preferred when compatible with limited information. This specific guideline is very important in terms of the approach to be taken in the risk assessment and it enters into the risk management appraisal later.

However, one cannot effectively discriminate among various mathematical models based upon how well they fit experimental observations. In the absence of adequate indications to the contrary, the linearized multistaged procedure must be utilized (per EPA).[6] The linearized multistaged model (discussed later in this chapter) leads to a plausible upper limit to the risk that is consistent with some proposed mechanisms of carcinogenesis; however, such an estimate does not necessarily give a realistic prediction of actual risk.

The true value of the risk is *unknown* and may be as low as zero! Therefore, the range of risk defined by the upper limit given by the chosen model and the lower limit (which may be as low as zero) should be explicitly stated in the approach.

Comparison

Low dose risk estimates derived from lab animal data extrapolated to humans are complicated by a variety of factors that differ among species and potentially affect the response to carcinogens. These items include life span, sex, body size, genetic variability, existence of other disease, exposure regimen, population homogeneity, and factors such as metabolism and excretion patterns.

The usual approach for making interspecies comparisons has been to use standardized scaling factors. Commonly employed standardized dosage scales include mg/kg body weight, ppm in the diet or water, mg/m^2 body surface area per day, and mg/kg body weight per lifetime. These scaling factors are related to the type of exposure pathway as well as the sensitivity of the various organs that may be involved.

Exposure Assessment

In order to obtain a quantitative estimate of carcinogen risk, the result of the dose-response assessment must be combined with an estimate of exposures to which the population of interest is likely to be subject. The exposure assessment is presented in a separate EPA guideline document.[4] Table 3.2 outlines the major elements required when performing an exposure assessment.

Table 3.2 Major Elements for an Exposure Assessment

1. Information for each chemical or mixture
    a. Identity of the substance(s)
    b. Chemical and physical properties

2. Sources
    a. Characterization of production and distribution
    b. Uses
    c. Disposal
    d. Summary of environmental releases

3. Exposure pathways and environmental fate
    a. Transport and transformation
    b. Identification of principal pathways of exposure
    c. Predicting environmental distribution

4. Measured or estimated concentrations
    a. Uses of measurements
    b. Estimation of environmental concentrations

5. Exposed populations
    a. Human populations
    b. Nonhuman populations

6. Integrated exposure analysis
    a. Calculations of exposure
        (1) Exposed population
        (2) Pathways of exposure
    b. Human dosimetry and biological measurements
    c. Development of exposure scenarios and profiles
    d. Evaluation of uncertainty
        (1) Limited initial data
        (2) Subjective estimates of input variables
        (3) Model input variables
        (4) Exposure

Source: U.S. EPA Office of Research and Development, Risk Assessment Guidelines, *Federal Register* (September 24, 1986).

These elements consist of the information on the specific chemical or mixture, which includes identity of the substance and its chemical and physical properties. The sources of the agent, its uses, disposal, characterization, production, and distribution, as well as a summary of the potential environmental releases for any given scenario, must be a part of an exposure assessment. The exposure pathways and the environmental fate, including transportation, transformations of the agent, and predictions of environmental distribution, must also be identified for each agent.

The measured or estimated concentrations to be used for the exposure assessment will include the types of measurements, how they are generated, and an estimation of the environmental concentrations that could be expected. This may involve some form of mathematical modeling of emissions from either a point source, such as a stack, or an area source, such as a storage pile or landfill. Calculations of exposed populations, intake, and the uncertainties involved in the preceding calculations round out the exposure assessment. Background concentrations must also be presented and taken into account.

Risk Characterization

Characterization of risk is composed of two parts: (1) a presentation of the numerical estimates of risk, and (2) a framework to help judge the significance of that risk. This framework consists of the exposure assessment and the dose-response assessment, and may also include a unit risk estimate. The latter can be combined with the exposure assessment for the purposes of estimating the cancer risk. Figure 3.1 shows, in a slightly different format, the major elements in risk characterization.

The numerical risk estimates can be presented in one of three ways:

• unit risk
• dose corresponding to given risk
• individual probability

Unit risk is the excess lifetime cancer risk attributable to a continuous lifetime exposure to "one unit" of carcinogen concentration under an assumption of low dose linearity. The second estimate is presented as the dose corresponding to a given level of risk (i.e, the dose corresponding to one excess cancer per million population exposed). This approach is used for nonlinear extrapolation models where the unit risk could differ at different dose levels. The third estimate is based upon individuals or populations (i.e., the probability of an individual or a group getting cancer for a given scenario).

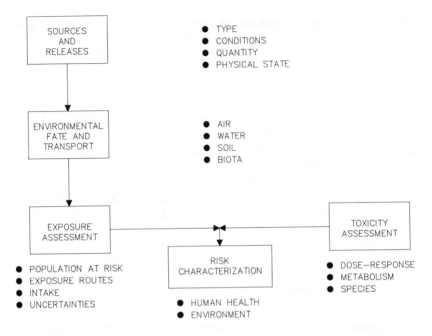

Figure 3.1.   The key features in a health risk assessment.

Irrespective of the option chosen, the degree of precision and accuracy in the numerical risk estimate currently does not permit more than one significant figure to be used.

In characterizing the risk due to concurrent exposure to several carcinogens, the risks are to be combined on the basis of additivity unless there is specific information to the contrary. Interactions of cocarcinogens, promoters, and initiators with other known carcinogens must be considered on a case-by-case basis.

Some of the major items which should be considered in risk characterizations are:

- the statistical uncertainties in estimating the extent of health effects
- how the uncertainties are to be computed and presented
- the biologic uncertainties in estimating the extent of health effects
- how they will be estimated
- the effect uncertainties will have on quantitative estimates

- how the uncertainties will be described to decision makers
- which dose-response assessments and exposure assessments will be used
- which population groups would be the primary target for protection

With respect to the latter, it should be noted that populations at particular risk include the elderly, the young, and those with preexisting health problems. Therefore, uncertainties on how they would respond to a given carcinogen or dose are critical compared to the general population at large.

Health Risk Assessment Requirements
for Hazardous Wastes

The 1984 HSWA amendments to the Resource Conservation and Recovery Act (RCRA)[7] require very specific information in a health risk assessment for an operator of a treatment, storage, and disposal hazardous waste management facility (TSDF). These requirements set the standard for what is required for any hazardous waste management activity, and are seen in Table 3.3. The three major categories are (1) the definition of a health assessment and how it pertains to risk, (2) the exposure assessment information, and (3) the health risk assessment based upon numerical calculation.

UNCERTAINTIES IN RISK ASSESSMENTS

There are two major categories of uncertainties: the toxicity of a chemical agent for humans and the exposure of a population group to that potential agent.

Uncertainties in Toxicity

Some uncertainties involve the testing scheme that was used in generating toxicity data for a given chemical for a given species. Others involve the potential synergism or antagonism that may occur between the chemical being observed and other coexisting chemicals or agents. It should be noted that tests are not an absolute measure of toxicity.

Table 3.3 Health Risk Assessment Requirements for a TSDF

1. **Definition of health assessment**
   - Nature and extent of contamination
   - Existence of potential for pathways of human exposure:
     - groundwater contamination
     - surface water contamination
     - air emissions
     - food chain contamination
   - Size and potential susceptibility of the population exposed
   - Comparison of expected *human* exposure levels to:
     - short-term health effects
     - long-term health effects
     - tolerance limits
     - existing data
   - Risk to affected population from all sources of contamination

2. **Exposure assessment information**
   - Reasonably foreseeable *potential* releases
     - normal operations
     - accidents
   - Potential pathways of *human* exposure
   - Potential magnitude and nature of human exposure

3. **Health assessment**
   - Required if a substantial potential risk exists

Source: Public Law 98–616, Hazardous and Solid Waste Act, Section 3019, Amendments to RCRA (PL 94–580) (1984).

Whether or not a threshold can be presumed to exist is also important. Recall that by policy, the EPA considers carcinogens to have no acceptable threshold of exposure. If such a threshold exists, it would indicate that the organism could detoxify some level of chemical exposure (which we know is the case for a large number of contaminants). If no threshold is assumed, this would indicate that some risk will always exist for any exposure to any level of a given chemical.

Regulatory agencies differ widely in their perception, understanding, and application of these principles. A case in point is the 1958 Delaney Amendment[8] to the Federal Food, Drug and Cosmetic Act. This law requires any chemical that causes cancer in lab animals to be banned from use in foods. Saccharin is one of these chemicals and yet it is still widely used today. This is due to a later congressional response that the perceived benefit of utilization of a noncaloric sweetener outweighs its potential risks.

Extrapolations from high dose effects to low dose effects are another major area of uncertainty with errors potentially ranging from 10^5 to 10^6 times. Interspecies differences and conversions of animal test data to humans are other sources of uncertainty involved in toxicity measurements. These two together will give uncertainties in the range of "orders of magnitude" (leading from 100 to 1,000 times differences in risk estimates).

Conversion of animal data for application to humans may by itself lead from 100 to 10,000 times uncertainties in final calculations.

Some factors of uncertainty can be in the range from 10 to 100 times. The purity of test compounds used in toxicity testing is one area that may introduce uncertainties in this range. Other uncertainties in this range include the laboratory procedures, the mechanics of obtaining a test sample, or the potential effect of mixtures of chemicals unknown to the researcher.

Some of the factors that result in uncertainties from 1 to 10 times include the purity of the strain of the test animal or potential interferences from disease or recessive genes.

Uncertainties in Modeled Exposures

Much of the data used for risk management decisions on proposed projects potentially emitting toxic or hazardous substances involve some form of mathematical modeling. There are therefore serious concerns as to the degree to which a mathematical model is able to accurately determine the concentrations to which one may be exposed.

Typically, mathematical dispersion models for air pollution are accurate to within a factor of ± 50–100%. Emission estimates (source terms) for point sources of criteria pollutants are generally accurate to within $\pm 20\%$.

For trace toxic emissions estimates, quite a different picture emerges. A recent publication by the South Coast Air Quality Management District (SCAQMD) compares the measured ambient air concentrations of organic compounds and trace toxic metals in the atmosphere with those predicted by models.[9] A summary is given in Table 3.4. The ratio of

Table 3.4 Comparison of Measured and Model-Predicted Toxic Air Pollutant Levels

Air Toxic	Measured	Model Predicted	Predicted/ Measured Ratio
Organic gases (ppb)			
Benzene	1.0–4.9	0.56–5.0	0.22–1.8
Carbon tetrachloride	0.10–0.12	$1.1–24 \times 10^{-5}$	$1.0–25 \times 10^{-4}$
Chloroform	0.02–0.30	$2.0–17 \times 10^{-8}$	$0.68–49 \times 10^{-7}$
Perchloroethylene	0.5–3.1	0.28–2.4	0.22–1.5
Trichloroethylene	1.1–7.1	0.33–2.9	0.13–54
Trace metals (ng/m³)			
Beryllium	0–0.5	$0–5.4 \times 10^{-3}$	0.003–3.4
Cadmium	0–4.1	1.1–9.6	0.71–1200
Chromium	1.8–11	3.6–60	1.06–8.6
Lead	180–280	1100–1700	3.9–9.4

Source: South Coast Air Quality Management District, 1987.

predicted to measured values for each of these contaminants is seen to vary from 0.68×10^{-7} for chloroform to 1,200 for trace metals such as cadmium.

A recent study by the Rand Corporation incorporated a comparison of health risk assessment models. In this study the same estimated arsenic emissions data were taken for a hypothetical coal-fired power plant and the risk calculated using four different models. The additional human lifetime cancer rate using the same data ranged from 0.0038 (probit dose-response model) to 0.33 (Weibull dose-response model).[10] This indicates a variation of 100 times just from using different models.

CALIFORNIA APPROACHES

Two agencies in California have developed their own approaches to risk assessment. The Department of Health Services (DHS) has produced the Site Mitigation Decision tree,[11] while the California Air Resources Board (CARB) has published its own document, the Air Toxics Source Assessment Manual,[12] for California air pollution control districts. Each of these is evaluated separately.

The Site Mitigation Decision Tree

This approach is notable in that, while it deals primarily with hazardous waste sites, it has an exhaustive method for identifying risks, addresses toxicological effects, and presents the toxicological models for determining the cancer risk rate. From there it develops an applied action level (AAL) that is used as a numerical criterion to be compared against modeled assessment of risk. Again, the end point considered for response is carcinogenicity. The section on developing risk factors for animal studies by low dose extrapolation using mathematical models is excellent. It provides a thorough evaluation of each of the five mathematical models to be used with the "decision tree" approach. The health risk assessment models and their explanations are given below.

The Linearized Multistaged Model

The linearized multistaged model is based on the following assumptions: (1) carcinogenesis results from an irreversible and self-replicating process that involves a number of different random biological events; (2) cancer originates in a single cell; (3) the time rate of occurrence of each event is in strict linear proportion to the rate of exposure or dose; and (4) the incidences of induced cancer and background cancer as measured in the control population are additive rather than independent of each other.

The mathematical expression for the linearized multistaged model is given by the equation:

$$P(d) = 1 - \exp[-(q_o + qd + \ldots q_k d^k)] \tag{1}$$

where $P(d)$ = the lifetime risk of cancer at the particular dose d; q_k = coefficients calculated to give the best fit to the experimental data; and k = the number of stages to be considered in fitting the model. For purposes of calculating $P(d)$, k may be set equal to N - 1, where N = the number of experimental dose groups. N is usually between 2 and 10. For k = 1, the linearized multistaged model reduces to the one-hit model originally used by EPA to derive its ambient water quality criteria.

The incremental or additional risk above background as measured in the experimental control population is given by the expression:

$$A(d) = [P(d) - P(0)] / [1 - P(0)] \tag{2}$$

where a(d) = the conditional probability of acquiring cancer at dose d, given that no cancer would have been detected in the absence of the dose; P(d) = the lifetime risk of cancer at dose d; and P(0) = the lifetime risk of cancer when d = (0) (i.e., background risk). For the multistaged model, substituting equation (1) into equation (2) gives:

$$A(d) = 1 - \exp[-(q_1 d + \ldots q_k d^k)] \qquad (3)$$

or, at low doses:

$$A(d) \cong q_1 d \qquad (4)$$

The parameter q_1 is approximated by the slope of a graph A(d) vs d (additional risk vs dose) at low doses.

The upper confidence level estimates of risk derived from this model are linear at low doses because the linear term of the model's mathematical expression dominates the other (nonlinear) terms at low doses.

The linearized multistaged model assumes that the dose-response curve has a positive, nonzero slope at zero dose. The model rules out the calculation of any dose below which carcinogenesis would not occur. Therefore, adoption of this model results in rejection of a threshold concept. The model also predicts that, at low doses, the number of expected additional cancers is proportional to the dose and that the age distribution of additional cancer incidence is independent of the dose rate. In other words, decreasing the dose would decrease the number of additional cancers at each age, but it would not change the relative distribution of cancers in the population age groups.

The One-Hit Model

The one-hit model, which was employed by EPA to develop the 1980 ambient water quality criteria, can be derived as a special case of the multistaged model by setting the number of stages equal to one. The one-hit model assumes that cancer incidence follows a Poisson distribution, that each additional molecule produces an equally probable response, and that all hits are independent of each other.

The probability of the occurrence of cancer is given by the equation:

$$P(d) = 1 - \exp[-(a + bd)] \qquad (5)$$

where a = background incidence of cancer and b = the slope of the dose-response curve. The slope of the dose-response curve is related to the potency of the carcinogen. At low doses the equation simplifies to:

$$P(d) \cong a + bd \qquad (6)$$

This model, like the linearized multistaged model, does not allow for a threshold effect. The model is quite conservative in that it results in higher risk estimates than most other models.[11]

The Gamma Multi-Hit Model

The gamma multi-hit model assumes that more than one hit or discrete event must occur to produce cancer. The response depends on the number of times the cell receptor is hit and on the potency of the agent.

The model also assumes a Poisson distribution of the number of hits over a fixed time period. The mathematical expression for the gamma multi-hit model is given by the equation:

$$P(d) = \sum_{i=K}^{\infty} (\Theta d)^i \exp(-\Theta d)/i! \qquad (7)$$

$$= \int_{0}^{\Theta d} [t^{k-1} \exp(-t)/(k-1)!] \, dt \qquad (8)$$

where $P(d)$ = the probability of response at dose d; k = the number of hits which result in a response; Θd = the mean for dose d of the Poisson distribution of the total number of hits; and t = the fixed time period. At low doses this may be approximated by the equation:

$$P(d) \cong (\Theta d)^k/\Gamma(k) \qquad (9)$$

$$\text{and} \quad \Gamma(k) = \int_{0}^{\infty} t^{k-1} \exp(-t) dt \qquad (10)$$

For $k > 1$, the estimated risks are between those estimated by the multistaged model and the log-probit model, a model discussed below. For $k = 1$, the gamma multi-hit model reduces to the one-hit model.

The Log-Probit Model

This model does not assume any biological carcinogenic mechanism, but rather rests on statistical assumptions. The log-probit model assumes that an individual tolerance level (threshold) exists such that a certain dose will not produce a response within the lifetime of that individual. The model also assumes that the logarithm of the individual tolerance doses is distributed normally (i.e, Gaussian distribution) in a population. The mathematical expression of the log-probit model is given by the equation:

$$P(d) = \phi (a + b \log d) \qquad (11)$$

where P(d) = the probability of a response at dose d; ϕ = the normal cumulative distribution function; a = the background incidence of cancer; and b = the slope of the log-probit distribution.

The Mantel-Bryan procedure,[11] a modification of the log-probit model, calculates the 99% confidence level of the risk associated with the highest no-response dose. The dose that corresponds to a predetermined risk level (e.g., 10^{-6}) is then derived. This dose is recommended as representing an acceptable risk.

Mantel and Bryan recommended using b = 1 to give a conservative (i.e., higher) risk estimate for most substances.[11] However, risk estimates using this method are still generally lower than those of the other models at low doses because the slope of the log-probit curve is comparatively steep.

Time-to-Response Models

Although most experimental data available measure dichotomous response, experiments may measure a time-to-response (e.g., the time from initial exposure to the first appearance of a tumor or the occurrence of death). These data can be employed to sort out other risks, such as acute toxicity, that could affect the carcinogenesis bioassay. Several low-dose extrapolation models can incorporate time-to-response data.

Druckrey[11] reported an empirical relationship between dose and time to tumor's appearance by analyzing data from studies which employed over 10,000 animals. Those data were described by the expression:

$$I = bt^k \qquad (12)$$

where I = the incidence of tumors at age t; and b and k are empirical constants.[11]

The Weibull model can be employed to predict response as a function of time and dose. This model can be described by the mathematical expression:

$$P(t,d) = 1 - \exp [g(d)t^k] \qquad (13)$$

where $P(t,d)$ = the probability of response at time t for a dose d; $g(d)$ = some function of dose d (e.g., $g(d) = a + bd^m$); t = the age of the animal at the end of the experiment; and k = the number of stages in carcinogenesis.[11]

Air Toxics Source Assessment Manual

In the CARB Guidance Documents for air toxics, a "cookbook" procedure is presented that allows for a screening of risks and then goes to a formal assessment procedure.[12] The screening level procedure provides for five separate activities, as does the formal assessment procedure. These activities are:

1. description of emission rates
2. estimation of ambient air concentrations
3. description of exposed populations
4. calculations of exposure from noninhalation pathways
5. the health risk assessment

The purpose of the screening technique is to facilitate the permit evaluation process for a potential source of air toxic emissions by making simplified assumptions that ensure compliance.

If a proposed project passes the initial screening elements, the air pollution control district and the public can be assured that this project does not pose a significant health risk to the community. In the event the potential project fails the screening process, the more formal assessment procedures would be required.

The formal assessment provides a more detailed estimate of the potential health risks. This approach by the CARB does not specify any particular cancer model. However, it does provide references to atmospheric dispersion modeling that may be used for calculating exposure. Rather than a cancer guideline, this document utilizes unit risks for a number of substances that have been developed by EPA's Carcinogen Assessment Group (CAG) and the DHS.[12]

In this approach, the calculation of excess lifetime cancer risk is determined by multiplying the ambient air concentration (determined by dispersion modeling) by the unit risk value times the fraction of an individual's lifetime for which the exposure occurs. This gives a numerical value for excess lifetime cancer risk. The CARB procedure, therefore, stops at the point of providing a numerical value of excess cancer risk.

It is notable, however, that this guideline provides for two cases: the worst case estimate and the most plausible case estimate of cancer risk rate. Also, noncancer effects are mentioned.

Under the formal health assessment, the worst case is made of assumptions that are conservative in every respect, i.e., worst emissions, worst meteorological conditions, and less-than-optimum performance of emission control devices. The most plausible scenario is that which could be expected if the emission rates were fairly well defined under average ambient meteorological conditions and optimum operation of air pollution control equipment.

CRITERIA FOR HEALTH IMPACT AND RISK

When noncarcinogenic risk is being evaluated, one is generally directed to using other published, acceptable levels of exposures. Some were alluded to earlier and include the TLVs modified by a number of factors ranging from 1/10 to 1/1,000. Since the TLVs were developed for health effects on healthy adult males with no more than 40 hours of exposure per week, these factors are assumed to provide a margin of safety for the general population. In addition, the PEL may be used in some cases.[13]

Where noninhalation pathways exist, the acceptable daily intake (ADI) may be used.[4,5] This is an EPA reference dose but it applies to intake of elements and toxic substances by the food and drink pathway only.

Where no published guidelines exist for comparison, the standard approach is to compare modeled concentrations to the ambient concentrations currently found in the area of concern. If the estimated concentration is less than 1% of the existing background, it is considered an acceptable intake or exposure. However, where the modeled exposure concentrations are in the same range or exceed the current average intake or background of a contaminant, guidelines are truly lacking. At this point, most agencies require a project proponent to suggest his own approach, either by analogy or by reference to other models or approaches.

When the perceived risk is large, risk management for a government agency is much different from that of a project proponent or a generator of hazardous substances.

For a government agency with vocal public constituencies, there are few options available apart from permit denial or a demand for immediate remedial action. (Here the difference between perceived risk and calculated risk is critical.) At this point, normally a decision or an action is taken as the end result of the risk assessment process.

For private concerns, the options are few, apart from early involvement of the public in the process, education of the regulatory personnel that may be involved, and providing for public participation. Other approaches, especially in siting of industrial facilities, have included negotiation, mediation, economic incentives, technical assistance, and local input into decision making. While these will not affect the scientific aspects of the risk assessment, they will certainly help with the perception of risk, which is the criteria by which most final decisions are made.

Internal risk management for a generator/emitter of hazardous substances generally takes the form of risk reduction through loss control techniques, waste minimization, or zero discharge policies. These are dealt with in later chapters.

REFERENCES

1. "Chemical Risk, A Primer," Information Pamphlet, American Chemical Society, Office of Federal Regulatory Programs/Department of Government Relations and Science Policy, Washington, DC (1984), p. 9.
2. "Transportation Fatalities (preliminary) — 1986 Data." AOPA 1988 Aviation Fact Card, Aircraft Owners and Pilots Association (Frederick, MD), p. 2.
3. Allman, W. F. "We Have Nothing To Fear," *Science*, October 1985, p. 39.
4. "Carcinogen Risk Assessment, Estimating Exposures, Mutagenicity Risk Assessment, Health Assessment of Suspect Development Toxicants, Health Risk Assessments of Chemical Mixtures." U.S. EPA Office of Research and Development, Risk Assessment Guidelines, *Federal Register*, September 24, 1986.
5. "Risk Assessment and Management: Framework for Decision Making," U.S. EPA Report 600/9-85-002 (1984), p. 11.
6. "Guidelines for Carcinogen Risk Assessment," *Federal Register*, September 24, 1986, p. 33997.
7. Public Law 98-616, 98 Stat. 3221 (1984) from HR 2867, 98th Congress, 2nd Session.
8. 21 USC, Section 348 (c) (3) (A).
9. "The Magnitude of Ambient Air Toxics Impact from Existing

Sources in the South Coast Air Basin," South Coast Air Quality Management District, 1987 Air Quality Management Plan Revision, Working Paper No. 3, Planning Division.

10. Boelten, J., P. Morrison, K. Soloman, and K. Wolf. "Alternative Models for Risk Assessment of Toxic Emissions," Rand Corporation, Report No. N2261-EPRI (1985).

11. "The California Site Mitigation Decision Tree: A Draft Working Document," California Department of Health Services, Toxics Substances Control Division, Alternative Technology and Policy Development Section (1985).

12. "Air Toxics Assessment Manual," California Air Pollution Control Officers Association (1987).

13. Hallenbeck, W. H., and K. M. Cunningham. *Quantitative Risk Assessment for Environmental and Occupational Health* (Chelsea, MI: Lewis Publishers, Inc., 1986).

4   AIR POLLUTION AND
AIR TOXICS

INTRODUCTION

An important exposure route to hazardous materials is by air. The health risks are apparent, but it is necessary to look at them in greater depth.

The subject of air toxics is a portion of the general field of air pollution just as the subject of hazardous waste is a portion of the general field of solid waste. In order to understand the field of air toxics, therefore, an initial discussion must focus on the generalized field of air pollution and air pollution control. In this section we will consider the general field of air contaminant emissions and air standards and then consider the regulations, agencies involved, and specific information on air toxics.

There are two broad regulatory classifications of air contaminants. These are the "criteria" pollutants (those for which there are established federal ambient air quality standards) and the "noncriteria" pollutants (those that do not have an established federal ambient air standard).

The criteria air pollutants include the gases — oxides of nitrogen, ozone, sulfur dioxide, and carbon monoxide — and the solids: particulate matter and lead (as a particulate).[1]

The noncriteria pollutants therefore include practically every other compound or element that could have an impact on human health or the environment. These may be further classified into those that are regulated under the Federal Clean Air Act, Section 112 (National Emission Standards for Hazardous Air Pollutants, or NESHAPs)[2] or those that are "toxic" (those chemicals and elements that a governmental agency believes may have adverse health effects).

Definitions

It is helpful if we define some of the terms before we proceed. Ambient air quality standards (AAQS) are those concentrations of contaminants in air above which adverse health effects occur. In areas of poor air quality these contaminant levels are also used as goals for attainment either at the federal or state level. A listing of the federal standards is given in Table 4.1. A comparison to California ambient air quality standards is also provided.

Emission standards, on the other hand, are limits on either mass emission rates or concentrations of contaminants from mobile or stationary sources.

Air toxics have different definitions at the federal, state, and local air quality district level. Under Section 112 of the Clean Air Act, a hazardous air pollutant is an "air pollutant to which no ambient air quality standard is applicable and which causes or contributes to air pollution which may reasonably be anticipated to result in an increase in mortality or an increase in serious irreversible, or incapacitating reversible, illness."[2]

For the state of California, a toxic air contaminant means an "air

Table 4.1 Ambient Air Quality Standards*

Contaminant	California	Federal (Primary)
Ozone	0.10 ppm	0.12 ppm
Carbon monoxide	20 ppm 9 ppm (8 hrs)	35 ppm 9 ppm (8 hr)
Nitrogen dioxide	0.25 ppm	0.05 ppm (annual)
Sulfur dioxide	0.25 ppm	0.14 ppm (24 hr) 0.03 ppm (annual)
Particulate matter (PM-10)	50 μg/m^3 (24 hr) 30 μg/m^3 (annual)	150 μg/m^3 (24 hr) 50 μg/m^3 (annual)
Sulfates	25 μg/m^3 (24 hr)	—
Lead	1.5 μg/m^3 (30 day)	1.5 μg/m^3 (3 months)
Hydrogen sulfide	0.03 ppm	—
Vinyl chloride	0.010 ppm (24 hr)	—

*One hour average unless otherwise indicated.

pollutant which may cause or contribute to an increase in mortality or an increase in serious illness or which may pose a present or potential hazard to human health."[3] Pesticides are also regulated as toxic air contaminants.

At one local agency, toxic air contaminants have been recently defined in a very broad manner: "Toxic and potentially toxic air contaminants are substances identified or currently under review for possible identification as toxic air contaminants, or those categorized by the Environmental Protection Agency as carcinogens or any other materials determined to be potentially toxic."[4] Thus we see that toxic air contaminants have a wide variety of definitions, depending upon the regulatory agency involved.

Table 4.2 lists those chemicals or substances on which the EPA has made a decision with respect to hazard and that are, or may be, included under NESHAPs. Table 4.3 lists those contaminants that have already been studied and would be candidates for air toxic emission control in Southern California. A number of the chemicals on Table 4.3 show up under different listings as toxic air contaminants, hazardous constituents, or hazardous materials.

Table 4.4 shows the different classifications and numbers of suspected and known toxic substances. These range from the National Institute of Occupational Safety and Health (NIOSH) Registry of Toxic Effects of Chemical Substances (1979) at 40,000-plus,[5] to 180 known animal carcinogens, to approximately 80 substances that are listed as known probable human carcinogens by the International Agency for Research on Cancer (IARC).[6] So far, only seven compounds have been listed as NESHAP substances.[2]

Classifications

The physical state of an air pollutant must be considered: either as a gas or a "solid." (Liquid droplets or aerosols are considered solids under most regulations.) A further classification of the physical state of these pollutants would include whether they are "inert" (having long life spans in the ambient atmosphere) or "reactive" (those that react quickly in the atmosphere to form other substances). These physical state and reactivity classifications of air contaminants are important in terms of how they are controlled and transported, and where they have their maximum health impact.

Table 4.2. Status of EPA Toxic Air Pollutant Decisions

Substance	Regulatory Decisions*
Arsenic	NESHAP promulgated
Asbestos	NESHAP promulgated
Benzene	NESHAP promulgated
Beryllium	NESHAP promulgated
Cadmium	Intent to list
1,3-Butadiene	Intent to list
Carbon tetrachloride	Intent to list
Chlorobenzenes	Not to regulate
Chlorofluorocarbons 113	Not to regulate
Chloroform	Intent to list
Chromium	Intent to list
Coke oven emissions	Listed under Section 112, CAA
Dioxins	Detailed assessment
Epichlorohydrin	Not to regulate
Ethylene dichloride	Intent to list
Ethylene oxide	Intent to list
Formaldehyde	Detailed assessment
Manganese	Not to regulate
Mercury	NESHAP promulgated
Methyl chloroform	Not to regulate
Methylene chloride	Intent to list
Nickel	Not to regulate
Perchloroethylene	Intent to list
Phenol	Not to regulate
Polycyclic organic matter	Not to regulate
Radionuclides	NESHAP promulgated
Trichloroethylene	Intent to list
Toluene	Not to regulate
Vinyl chloride	NESHAP promulgated
Vinylidene chloride	Not to regulate

Source: U.S. EPA, Office of Air Quality Planning and Standards, 2/23/87.
*Regulatory decisions:
NESHAP promulgated—final rules have been issued.
Intent to list—preliminary health assessment complete and results indicate potential for significant health risk. Detailed evaluations have been initiated to identify sources warranting regulation.
Not to regulate—preliminary health assessment complete and results either indicate a lack of potential for significant health risk or a lack of health data. Federal regulation not warranted at this time.
Detailed assessment—data on emissions concentrations for sources of chemicals in the U.S. are in the process of being gathered. Health data base is being evaluated.

Table 4.3 Toxic Air Pollutants Studies in the South Coast Air Basin

Metals	Organics
Arsenic[1,2,3]	Benzene[1,2,3]
	Carbon tetrachloride[2,3]
Beryllium[1,2,3]	Chloroform[2,3]
	Ethylene dibromide[3]
	Ethylene dichloride[3]
Cadmium[2,3]	Methyl bromide[3]
	Methylene chloride[3]
Chromium[2,3]	Perchloroethylene[3]
	Toluene[2,3]
Lead[2,3]*	1,1,1 Trichloroethane[2]
Mercury[1,2,3]	Trichloroethylene[2,3]
	Vinyl chloride[1,2,3]
Nickel[2,3]	Xylenes[3]

[1]Hazardous air pollutants (NESHAP).
[2]EPA hazardous constituent (RCRA).
[3]California hazardous material.
*Lead is also a criteria air pollutant.

Table 4.4 Suspected and Known Toxic Substances

Category	Approximate Number of Substances
NIOSH registry of toxic effects of chemical substances	40,000*
IARC list of animal carcinogens with sufficient evidence of cacinogenicity	180
IARC list of known and probable human carcinogens (1982)	80
NESHAP substances	7
Substances proposed for NESHAP review	30
Potential toxic air contaminants	40–50

Source: South Coast Air Quality Management District, Air Quality Management Plan Revision, Working Paper No. 3 (1987).
*Plus 2000–5000 per year.

Sources

Mobile and Stationary Sources

The sources of normal air pollutants and toxic air contaminants fall into two broad categories: mobile and stationary. The mobile sources include contaminant-emitting engines that are usually associated with transportation, such as automobiles, trucks, trains, and ships. (Airplanes are generally associated with airports and so most of the impacts are associated with the airport area itself.)

Stationary sources include pipelines, factories, boilers, storage vessels, storage tanks, etc. Stationary sources are further classified into those that are point sources (chimneys, stacks, open vents, flares, etc.) and those that are area sources (parking lots, landfills, airports, and major industrial facilities).[7] Fugitive sources like waste piles, storage areas, loading areas, and transfer operations (such as conveyor belts) are a third category. Each of these is regulated in a different manner depending upon location, emissions, and the level of regulatory authority overseeing each operation.

Direct and Indirect Sources

Toxic air emission sources generally fall into two major categories: direct emissions to the atmosphere, and indirect emissions to the atmosphere.

Direct air toxic emissions are usually volatile in nature and come from either leaks, spills, or process vents. Toxic metals, as direct emissions, are not destroyed in any process but may remain present as ashes or particulates which escape control. In extreme cases, toxic metals may occur in the vapor phase of a high-temperature process and thus escape control also.

Indirect emissions are those formed by thermal processes.[8] These are usually combustion-derived gases and particles and depend on the type of fuel or the type of waste being burned. These air contaminants include aldehydes, carbon monoxide, dioxins, and benzene. The latter is an example of a product of incomplete combustion (PIC) of another toxic chemical called toluene. It is not unusual for combustion sources such as flares (which have both materials originally present in the fuel gas) to actually show a greater amount of benzene coming out of the combustion process than going in. This is a result of toluene being partially burned and then forming benzene in the exhaust.

Table 4.5 provides a compilation of the major emissions by point, area, and mobile source of the various air toxics in the South Coast Air Basin (California) as inventoried in 1984. Fuel-related emissions (lead, toluene, xylene) are directly related to mobile sources, whereas solvent emissions (1,1,1-TCA, vinyl chloride, methyl bromide) are related to stationary sources such as degreasing equipment, landfills, and fumigation practices.

It must be remembered that even air pollution control devices themselves (fabric filters, wet scrubbers) may become sources of hazardous waste, even though the exhaust air is being cleaned.[9] The residues collected from the exhaust gases may contain trace toxic metals or adsorbed organic compounds such as dioxins, which may pose a risk to public health. Thus, a complete evaluation of all exhaust streams leaving an air pollution control system is necessary.

Table 4.5 Emissions of Hazardous Materials in the South Coast Air Basin

Species	Emissions (tons/year)			
	Point	Area	Mobile	Total
Arsenic	0.047	–	–	0.047
Benzene	118.	7,870.	6,910.	14,898.
Cadmium	1.12	–	6.91	8.03
Carbon tetrachloride	3.20	–	–	3.20
Chromium	16.0	–	13.2	29.2
Lead	14.5	–	2,030.	2,045.
Mercury	0.13	–	–	0.13
Methyl bromide	24.4	–	–	24.4
Methylene chloride	4,780.	10,200.	–	14,980.
Nickel	5.40	–	2.44	7.84
Perchloroethylene	3,970.	8,850.	–	12,820.
Toluene	714.	276.	14,200.	15,190.
1,1,1-Trichloroethane	8,590.	6,150.	–	14,740.
Vinyl chloride	1.37	–	–	1.37
Xylenes	230.	185.	8,950.	9,365.

Source: South Coast AQMD, 1984.
– = no data available.

REGULATORY APPROACHES

The federal government has primary authority for regulating emissions from mobile sources. California, however, has a special Congressional dispensation to require more stringent emission limits from automobiles and trucks.[10]

The emission rates that are allowed for classes of mobile sources are set by the EPA, so manufacturers design and build vehicles and engines to meet those standards. Mobile sources emit far more gaseous emissions than particulate emissions.

Stationary sources of air contaminant emissions are regulated by local authorities under the jurisdiction of the respective state. Inspection and maintenance of vehicles for air emissions are also regulated under many state laws. The exception is for new major stationary sources of air pollution (such as power plants), which must also meet national standards. National stationary source standards are termed New Source Performance Standards (NSPS). These are emission standards adopted by the federal government on an industry-specific basis for all new sources of air contaminant-emitting equipment or processes located anywhere in the United States.[2]

The State Implementation Plan

The method by which states are to bring local ambient air quality into compliance with federal standards is termed the state implementation plan (SIP). The SIP is a document in which each state details to the federal government how, in what manner, and by which regulations the state will attain those air quality goals. This detailed plan demonstrates by emission calculations, new regulations, and planning efforts how the state will meet the air quality standards by a given year. The SIP, which will vary for different contaminants depending on the existing air quality in the local area, includes use of modeling, transportation control plans, and indirect source reviews as well as current strategies directed at specific stationary sources.

The local air pollution control and air quality management districts operate under the authority of the different states. The regulations adopted by the local districts spell out either performance standards (i.e., percent efficiency of a control device) or design standards for specific pieces of equipment (i.e., floating roofs on petroleum storage tanks to control volatile organic compounds).

The Permit System

The heart of the program for stationary source control by a local agency is the permit system. There are two types of permits (unless specifically exempted) to operate equipment:

1. permits for equipment that may cause air pollution
2. permits for equipment designed to control air pollution

Permits are required for new equipment, before modifying existing equipment, whenever a business changes ownership, or when the equipment is transferred from one physical location to another.

Permits are required for new or modified facilities before construction begins. Failure to obtain permits results in additional fees or civil or criminal penalties. The permits are thus the primary tool the district uses to ensure that businesses follow air pollution control laws. By granting a permit, the district indicates that equipment should be able to meet all the air pollution standards. Permits may also specify limitations or operating conditions for the equipment.

Typically, the permit process may involve two separate permits – a permit to construct and a permit to operate. Both permits may be issued under the same application.

For the permit to construct, the application must include all designs and engineering calculations needed for the proposed equipment or project. The engineering staff of the district then evaluates the application to make sure that by engineering standards it will be capable of meeting the district regulations. This allows the applicant to make changes in the design phase of a project rather than making later modifications to the completed structure. Unless the project is a "significant project," the facility is then authorized to begin construction. A significant project is any proposed plant emitting a toxic or potentially toxic air contaminant.

For a "significant project" in southern California, a public notice announcing a public hearing to inform of the potential air quality impacts of that proposed facility is required to be distributed to every address within a quarter-mile of the proposed project. Specifically named as significant is any plant that has to provide emission offsets or is a "refinery, a power plant, sewage treatment plant, or resource recovery/cogeneration facility."[4]

After construction, the operating equipment is evaluated by the district engineering and inspection personnel. Usually a source test is performed at the facility to ensure that equipment emitting air contaminants is operating in compliance with the law. Following that demonstration of compliance, a permit to operate is issued by the district. This then allows one to operate that facility or piece of equipment.

Federal Air Toxics Agencies and Their Roles

The EPA is the major federal agency involved in the regulation of ambient air toxics. Its primary involvement includes setting ambient air quality standards, establishing emission standards for new stationary and mobile sources, setting the NESHAPs, and enforcing Clean Air Act Section 112 (which includes air toxic studies and programs). The EPA also provides funding for a variety of other state and local agencies to carry out the overall enforcement and administrative programs.

The second agency charged with air toxic regulations is OSHA. This agency deals with exposures of individuals to potential toxic air contaminants while on the job. Operating under the Department of Labor, OSHA has tremendous authority when it comes to regulating exposures in the general workplace. This especially applies to workers at hazardous waste sites or locations where hazardous materials have been deposited, are currently being managed, or are currently being tested or treated.

OSHA's authority extends to actually shutting down an operation on the basis of an alleged or actual air toxic emission that could impact a worker's health. OSHA uses PELs as its air quality standards.

Air Toxic Regulations

The federal EPA regulations specifically for air toxics comprise the NESHAPs. These are required under authority of Section 112 of the Clean Air Act.[2] The NESHAP contaminants are usually regulated when they are associated with a specific type of equipment or operation. Thus, beryllium emissions are regulated for rocket motor firing but not for coal-fired power plants. The mercury NESHAP applies to chlor-alkali plants rather than emissions from other sources. The NESHAP for vinyl chloride applies to polyvinyl chloride manufacturing plants and the benzene NESHAP applies to fugitive equipment leaks rather than gasoline handling. The other federal regulations impacting air toxics, such as RCRA[11] and the Toxic Substances Control Act (TSCA),[12] are concerned not so much with regulating air emissions as quantifying them for risk assessment or worker exposure purposes.

There are two exceptions to this: the TSCA requirement for emissions from PCB incinerators and the RCRA requirements for incinerators burning hazardous wastes.

The TSCA air pollution control requirement is that an incinerator demonstrate a PCB destruction efficiency of 99.9999%.[13] Thus, of all of the PCBs going into the incinerator, no more than one gram in a million may be emitted.

The RCRA air pollution control requirements for hazardous waste incinerators are:

- for the principal organic hazardous constituents (POHCs): 99.99% Destruction and Removal Efficiency (DRE)
- hydrochloric acid removal: 99% removal or four pounds per hour (whichever is greater)
- particulate emissions: not exceeding 180 mg/m³ at 7% oxygen[14]

POHCs are the major toxic chemicals found in the particular hazardous waste stream being incinerated. These would include most of the air toxics identified in this chapter.

REFERENCES

1. Danielson, J. A. *Air Pollution Engineering Manual*, 2nd ed., U.S. EPA Office of Air Quality Planning and Standards, Pub. No. AP-40 (May 1973).
2. Public Law 95-95, Clean Air Act Amendments of 1977, Section 112 (1977).
3. California Health and Safety Code, Section 39655.
4. South Coast Air Quality Management District, Rule 212 (July 1987).
5. "Registry of Toxic Effects of Chemical Substances," National Institute for Occupational Safety and Health, U.S. Department of Health and Human Services (1979).
6. "IARC Monographs on the Evaluation of the Carcinogen Risk of Chemicals to Humans," International Agency for Research on Cancer, Lyon, France (1982).
7. Woods, J. A., and M. L. Porter. *Hazardous Pollutants in Class 2 Landfills,* Technical Services Division, Laboratory Services Branch, South Coast Air Quality Management District (December 1983).
8. Brunner, C. *Hazardous Air Emissions from Incineration* (New York: Chapman and Hall, 1985).

9. Calvert, S., and H. M. Englund, eds., *Handbook of Air Pollution Technology* (New York: John Wiley and Sons, Inc., 1984).
10. Public Law 95-95, Clean Air Act Amendments, Section 209 (b) (1) (1977).
11. Public Law 98-616, Hazardous and Solid Waste Amendments (1984).
12. Public Law 94-469, Toxic Substances Control Act (1976).
13. Public Law 94-469, Toxic Substances Control Act, Section 761.40 (1976).
14. 40 CFR 264.343.

5   GROUNDWATER

INTRODUCTION

Groundwater is a major natural resource and represents more than 95% of all available fresh water in the United States. More than one-half of the U.S. population uses groundwater for their drinking water supply, and approximately 75% of the major U.S. cities depend on groundwater for their potable water supply.

Unfortunately, groundwater inadvertently serves as a sink for wastes and consequently is a potential route of exposure to hazardous materials. Wastes from accidental spills, runoff of agricultural and domestic fertilizers, insecticides and pesticides, leaky sewers, salts from urban and highway storm water runoff, landfills, air pollutant emissions, and wastewater from residential septic tanks, cesspools, and industrial lagoons may migrate into groundwater, where, in effect, these wastes are stored. Figure 5.1 shows how this may occur from an air pollution source. In addition, direct disposal to groundwater is practiced through the use of underground injection wells where wastes are pumped into geologic formations containing water unsuitable for drinking purposes. Disposal through these deep wells is conducted nationwide for wastes such as steel pickling liquor and brine from oil and gas production. The actual extent of total groundwater contamination is unknown.

The most frequently occurring contaminants are industrial solvents, hydrocarbon or oil products, and heavy metals. These are distributed in two ways: area sources (those that cover a large area, e.g., pesticides from landfills) and point sources (those discharged from a single small location, e.g., a leaking underground tank).

In order to truly understand protection of groundwater resources, it is necessary that one be aware of the fundamentals of groundwater hydrology, contamination in groundwater, and the mitigation methods avail-

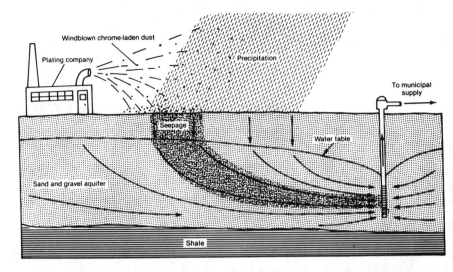

Figure 5.1.   An example of air pollution becoming a source of ground-water contamination.

able for cleanup of contaminated groundwater.[1,2] This chapter will review each of these major areas.

PRINCIPLES OF HYDROGEOLOGY

Groundwater exists almost everywhere. It occurs in saturated zones close to the surface of the earth, as well as thousands of feet below. To understand a groundwater system, one must understand the entire underground water regime, including the geologic formations in which it exists. A groundwater system is interactive, complex, and multifaceted. It is more than just a resource for human use; it is integral to the environment in many ways. For instance, the normal flow of major lakes and streams in the summer is supplied by groundwater.

Groundwater movement is not mysterious, but rather follows clearly defined principles of physics. The purpose of this section is to describe in a simplified manner (1) the physical characteristics of aquifers and how they function as storage media and as conduits, and (2) the principles of fluid flow.

Most of the rocks near the earth's surface are composed of both solids and voids, as Figure 5.2 shows. The solid part is, of course, more obvi-

Primary Openings

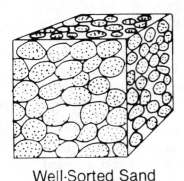

Well-Sorted Sand

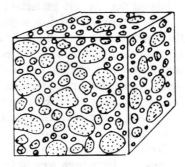

Poorly-Sorted Sand

Secondary Openings

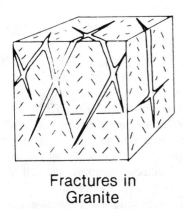

Fractures in
Granite

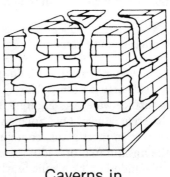

Caverns in
Limestone

Figure 5.2.   Primary and secondary openings in rocks.

ous than the voids, but without the voids, there would be no water storage or movement to supply wells and springs.

Water-bearing rocks consist either of unconsolidated (soil-like) deposits or consolidated rocks. The earth's surface in most places is formed by soil and by unconsolidated deposits that range in thickness from a few centimeters (near outcrops of consolidated rocks) to more than 12,000 meters (beneath the delta of the Mississippi River).[3] Unconsolidated deposits are underlain everywhere by consolidated rocks.

Most unconsolidated deposits consist of material derived from the disintegration of consolidated rocks. The material consists, in different types of unconsolidated deposits, of particles of rocks or minerals ranging in size from fractions of a millimeter (clay) to several meters (boulders). Unconsolidated deposits important in groundwater hydrology include, in order of increasing grain size, clay, silt, sand, and gravel. An important group of unconsolidated deposits also includes fragments of shells of marine organisms.

Consolidated rocks consist of mineral particles of different sizes and shapes that have been welded by heat and pressure or by chemical reactions into a solid mass. Such rocks are commonly referred to in geologic terms as bedrock. They include sedimentary rocks that were originally unconsolidated and igneous rocks formed from a molten state. Consolidated sedimentary rocks important in groundwater hydrology include limestone, dolomite, shale, siltstone, sandstone, and conglomerate. Igneous rocks include granite and basalt.

Underground Water

All water beneath the land surface is referred to as underground water (or subsurface water). The equivalent term for water on the land surface is surface water. As Figure 5.3 shows, underground water occurs in two different zones. One zone, which occurs immediately below the land surface in most areas, contains both water and air and is referred to as the unsaturated or vadose zone. The unsaturated zone is almost invariably underlain by a zone in which all interconnected openings are full of water. This zone is referred to as the saturated or phreatic zone.

Water in the phreatic zone is the only underground water that is available to supply wells and springs and is the only water to which the name groundwater is correctly applied. Recharge of the phreatic zone occurs by percolation of water from the land surface through the vadose zone. The vadose zone is, therefore, of great importance to groundwater hydrology.

The Hydrologic Cycle

Groundwater is a major feature of the hydrologic cycle, which is the water circulatory system for all the earth's fresh water. Water moves

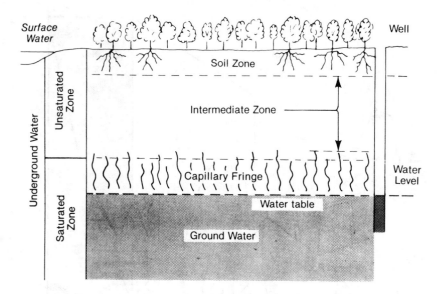

Figure 5.3. Groundwater occurs in the saturated zone.

through this cycle between the ocean, atmosphere, and land (Figure 5.4).

Water enters the land portion of the hydrologic cycle as precipitation: rain, snow, hail, and sleet. Part of this precipitation returns to the atmosphere through evaporation. Some becomes runoff, flowing via streams and rivers to lakes, other surface reservoirs, and oceans. The rest enters the soil. This water may be absorbed by plants where it may again enter the atmosphere through a process known as transpiration. The excess precipitation (beyond that removed by runoff, evaporation, and transpiration) percolates down through the earth. It will pass through a zone of aeration to the saturated zone.

The aerated zone is usually not saturated with water, except temporarily after a significant rainfall. This zone is composed of pore spaces, unevenly occupied by air and a small amount of water. Water contained here is held by surface tension, which occurs between the water and the particles of soil. This aerated zone typically extends from the ground surface past the root zone of plants.

Water that is not withdrawn by plants in the aerated zone continues to move slowly down to another zone where the pore spaces are saturated. The surface of the saturated zone is called the water table. Here, at the water table, and below, all the openings are filled with water (hence the

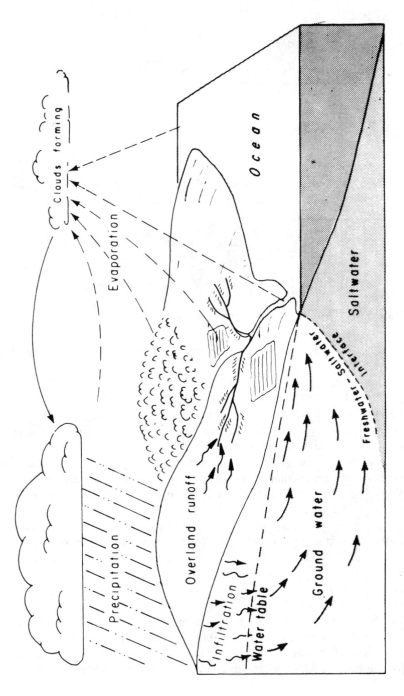

Figure 5.4.   The hydrologic cycle.

name "saturated zone") and the pressure is greater than the atmospheric pressure. Although "boundary" is a useful metaphor for describing the water table, precise lines between the aerated and saturated zones are not often found in nature.

Aquifers

The saturated zone is more frequently referred to as an aquifer. Although "aquifer" can mean many things, it is best defined as an underground, saturated, permeable geologic formation capable of providing significant quantities of water to a well or spring. It is the ability of the saturated zone (or a portion of that zone) to yield water that makes it an aquifer. Aquifers represent the usable portions of groundwater, and are, in effect, underground storage areas or reservoirs.

Aquifers are commonly classified by the geologic material of which they are composed. Seldom homogeneous in composition, aquifers may well be a mixture of several materials. Unconsolidated sands and gravels compose nearly 90% of all aquifers developed for water supplies. Porous sandstone, limestone, and highly fractured crystalline and volcanic rock are also common aquifer materials. A saturated zone, or portion of it, that is of low permeability and will *not* provide significant quantities of water to a well or spring is referred to as an aquitard. Dense crystalline rock, clays, and shales are common aquitards. Water-bearing layers can be found on top of each other (aquifer-aquifer layers) or in aquifer-aquitard combinations.

Aquifers can range in size from a narrow strip of gravel adjacent to a river to regional aquifers covering hundreds of thousands of square miles. The zone where water enters an aquifer is called the recharge area. It enters as precipitation or infiltration from surface waters (e.g., overlying streams and lakes), flowing through the pore spaces at an imperceptible but nonetheless unceasing pace. Groundwater velocities vary widely, ranging from several millimeters to several feet per day. The permeability of the aquifer will depend on various physical factors such as the size and degree of interconnections among the pore spaces, as well as the fluid's properties. A highly permeable water-bearing layer will have large particles with many interconnected pore spaces large enough for the liquid to migrate freely.

Water in an aquifer will move downhill under the influence of the hydraulic gradient. This gradient may be considered the slope of the water table. Sometimes groundwater will appear to move uphill. This is because it is under hydraulic pressure as a result of more water moving downhill, forcing other water up. The rate of groundwater flow depends on permeability and hydraulic gradient. Groundwater flows through the aquifer until it emerges in a discharge area (e.g., well, lake, spring, geyser, river, ocean). The age of the groundwater (the time spent underground) will vary according to its velocity and the distance between recharge and discharge areas.

Two types of aquifer exist: unconfined and confined. An unconfined aquifer is near the earth's surface and is easily recharged. It also is known as a water-table aquifer because the water table is its upper boundary.

Confined aquifers occur farther below the ground surface. They are wedged between aquitards and are consequently under pressure from the overlying, relatively impermeable strata. These aquifers are often referred to as "artesian" or pressure aquifers. Many artesian aquifers are continuous for a long distance, becoming unconfined where the aquitard ends, usually at the earth's surface. Figure 5.5 illustrates the different types of aquifers and geologic strata.

Hydraulic Conductivity

Aquifers transmit water from recharge areas to discharge areas and thus function as porous conduits (or pipelines filled with sand or other water-bearing material). The factors controlling groundwater movement were first expressed in the form of an equation by Henri Darcy, a French engineer, in 1856.[4] One form of Darcy's Law is given as:

$$Q = KA \frac{dh}{dl}$$

where Q = the quantity of water flowing per unit of time; K = the hydraulic conductivity, which depends on the size and arrangement of the water-transmitting openings (pores and fractures) and on the characteristics of the fluid (water) such as kinematic viscosity, density, and the strength of the gravitational field; A = the cross-sectional area, at a right angle to the flow direction, through which the flow occurs; and dh/dl = the hydraulic gradient.

Because the quantity of water (Q) is directly proportional to the hydraulic gradient (dh/dl), we say that groundwater flow is laminar, that is, water particles tend to follow discrete streamlines and not to mix with

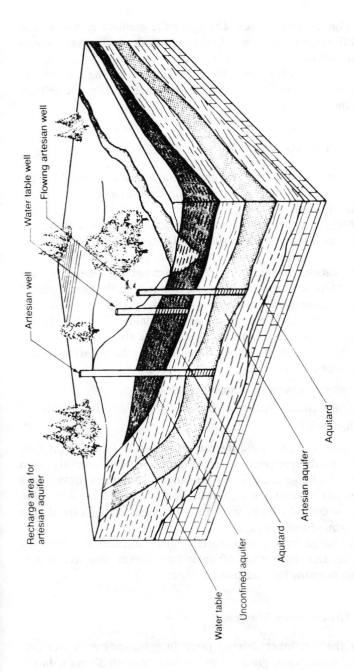

Recharge area for
artesian aquifer

Artesian well

Water table well

Flowing artesian well

Water table

Unconfined aquifer

Aquitard

Artesian aquifer

Aquitard

Figure 5.5.   Aquifers and geologic strata.

particles in adjacent streamlines. The hydraulic gradient is thus a measure of the difference in elevation of the groundwater between two points in the same aquifer.

Darcy's Law then makes possible a calculation of the quantity of groundwater flowing between two points, or more usually, the velocity of the groundwater flow.

GROUNDWATER MOVEMENT

It is important to understand that groundwater moves at speeds on the order of feet per year in a given direction. Surface waters move at velocities of feet per second.

Water enters groundwater systems in recharge areas and moves through them, as dictated by hydraulic gradients and hydraulic conductivities, to discharge areas.

The identification of recharge areas is important because of the expanded use of the land surface for waste disposal. In the humid part of the country, recharge occurs in all interstream areas (that is, in all areas except along streams and their adjoining floodplains). The streams and floodplains are, under most conditions, discharge areas.

The rate of movement of groundwater from recharge areas to discharge areas depends on:

• if water moves downward into other aquifers
• the hydraulic conductivities of the aquifers and confining beds
• hydraulic gradients

A convenient way of showing the rate of flow is in terms of the time required for groundwater to move from different parts of a recharge area to the nearest discharge area. The time ranges from a few days in the zone adjacent to the discharge area to thousands of years for water that moves from the central part to some recharge areas through the deeper parts of the groundwater system.

Natural discharge from groundwater systems includes not only the flow of springs and the seepage of water into stream channels or wetlands but also evaporation and transpiration.

Surface and Groundwater Movement

Gravity is the dominant driving force in groundwater movement. Under natural conditions, groundwater moves "downhill" until, in the course of its movement, it reaches the land surface at a spring or through a seep along the side or bottom of a stream channel or an estuary.

Thus, groundwater in the shallowest part of the saturated zone moves from interstream areas toward streams or the coast. The depth to the water table is greater along the divide between streams than it is beneath the floodplain. In effect, the water table usually is a subdued replica of the land surface.

In most areas, general but valuable conclusions about the direction of groundwater movement can be derived from observations of land-surface topography. Also, surface streams may be visualized as being underlain by shallow groundwater movement. However, specific local conditions may vary widely.

CONTAMINATION IN GROUNDWATER

To understand groundwater contamination, one must understand the flow pattern of chemical substances in porous materials. Entry into the subsurface environment is unique for each type of contaminant source. A point source is a single-location source of contamination such as a waste site or an underground storage tank. A diffuse source is a source of contamination occupying an extensive area that is not clearly defined, such as a large area sprayed with an agricultural chemical, or an old landfill.

Once in the saturated groundwater flow, contaminants will not disperse throughout the entire aquifer. Rather, they will move with the flow of groundwater, tending to form irregular shapes of contaminated water as they move through the system. This flowing body of contaminated water is called a plume, and extends from the contamination source to some downstream point. The size and shape of the plume depend upon many variables, some of which include local geology, regional ground-water flow, type and concentration of contaminants, degree of continuity of the contaminant source, and any modifications of the groundwater system, such as well-pumping.

Plumes tend to become stable in localized areas. Plume stability varies widely, based on a large number of parameters, only one of which is a nearly constant input of contaminants into the ground. Stability may reflect a very slow movement of water or attenuation mechanisms at work (the numerous interactions of contaminants and aquifer materials).

A stable plume is like a steady stream which expands or contracts in response to such factors as the changes in the rate of contaminant input. As contaminant discharge continues, the plume remains localized beneath the input area and extending principally downstream. If the input ceases, the plume moves away from the source as a contaminated mass, like a cloud, in the direction of the flow.

Distortions, however, can occur in the flow if the geologic formation is heterogeneous and possesses varying resistances. For example, contaminated plumes might spread or form "fingers" caused by layered beds of rock or more permeable unconsolidated materials.

When a mixture of contaminants enters the saturated zone, each chemical species generally will travel at its own velocity, although some contaminants will aid the transport of others. This situation creates a contaminant mass or cloud that is segregated into different zones, each advancing in the same direction, but at different rates. Figure 5.6 illustrates such a plume containing different contaminants. The rate of flow of groundwater may create a situation where mixing of contaminants is so slow that dilution processes are retarded to the extent that localized plumes can contain high concentrations of contaminants.

Since groundwater tends to move slowly, the flow is termed laminar. This means that very little mixing occurs between a dissolved liquid and the flowing groundwater around it. In general, the angle of dispersion of a contaminated plume in a homogenous medium is less then 10°.

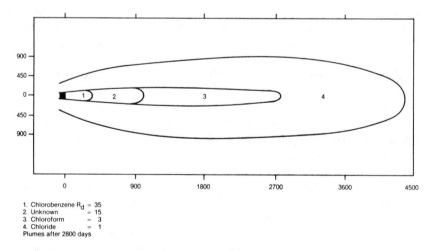

1. Chlorobenzene R_d = 35
2. Unknown = 15
3. Chloroform = 3
4. Chloride = 1
Plumes after 2800 days

Figure 5.6.   Differential movement of pollutants in a contaminated groundwater plume.

Dispersion is important in the study of groundwater pollution. However, it is difficult to measure in the field because the rate and direction of movement of contaminants are also affected by stratification, ion exchange, filtration, and other conditions and processes. Stratification, differences in lithology and other characteristics of aquifers and confining beds typically result in much greater lateral than vertical dispersion.

The attenuation rate of each contaminant, and the local precipitation and aquifer conditions (geologic strata and reservoir properties), affect the distance, direction, and rate of movement of each constituent in groundwater. Some of the important processes controlling the behavior of contaminants and modifying the characteristics of groundwater include dilution, filtration, adsorption, microbial activity, and chemical reactions.

Each group of contaminants is controlled by properties specific to its group and by subsurface conditions. Thus, groups of contaminants will behave differently. Some groups will be attenuated totally by specific processes, others not at all. For example, solubility constraints (derived from the ability to form precipitates with various minerals) tend to control the concentration of most nonmetallic, inorganic contaminants.

Mechanisms such as adsorption, biological degradation and uptake, and chemical degradation seem to prevent or retard the migration of some contaminants in the subsurface environment, often causing them to be left behind in a localized area. One example would be the adsorption of lead in alkaline clay soils.

Detection and Monitoring

Subsurface investigations are most often conducted with equipment that extends underground through monitoring or observation wells. From these wells, samples of water are collected to yield information on the quantity and quality of the groundwater, possible contaminants present, the distance traveled by these contaminants, and their direction. These investigations are being enhanced by geophysical techniques and by predictive computer models, such as those developed by the U.S. Geologic Survey (USGS).

Detection and monitoring of groundwater contamination requires a site-specific analysis. This analysis is complex because of the unique characteristics of subsurface waters which hamper effective monitoring.

In general, monitoring wells or well point probes are utilized to determine the extent of groundwater contamination beneath and downgradient ("downstream") of a contamination source. Other methods include soil gas analysis. In addition, "upgradient" wells are sampled to determine contamination levels before the groundwater encounters the plume. Vertical as well as lateral extent is important in installing monitoring wells, since contaminants may not follow the general flow direction.

Factors to Consider

The solubility of a contaminant as well as its density related to water is of particular importance both when placing wells for monitoring and in verifying location and flow patterns of contaminants. Water-soluble contaminants would tend to dissolve in and then flow with groundwater as seen in Figure 5.7. In Figure 5.8, however, we see the effect of another liquid (gasoline) encountering an aquifer after a spill. Here the gasoline, being relatively insoluble and less dense, tends to float on the surface of the aquifer. In Figure 5.9, however, we see the effect of a dense insoluble

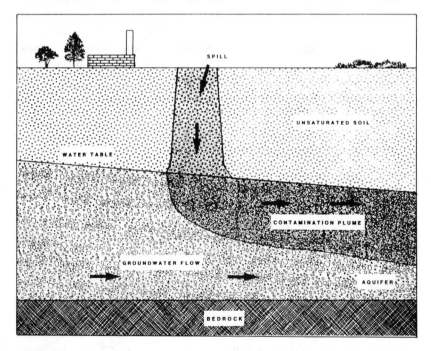

Figure 5.7.   Cross section of a dissolved contamination plume in an aquifer.

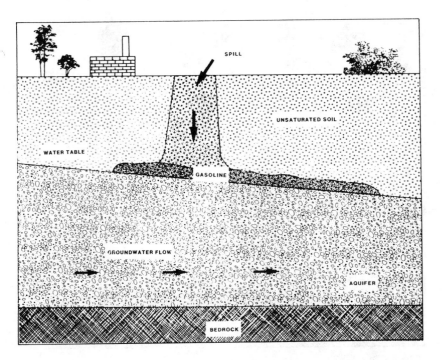

Figure 5.8. Cross section of a nonaqueous fluid less dense than water encountering an aquifer.

liquid encountering an aquifer. In this latter case the contaminant liquid goes to the bottom of the aquifer and thus might not be found in a well placed at the "proper" depth. And, in the case where a bedrock formation is tilted, it is possible for the contaminant to actually flow in a reverse direction to that of the groundwater!

Well materials and sampling methods are critical in gathering good quality data from groundwater. The effect of a contaminant such as MEK in groundwater on PVC pipe is such that (1) the pipe or well may be dissolved, or (2) the sample may be contaminated with dissolved materials such as vinyl chloride.

The general levels of groundwater contamination, especially for organic chemicals, may be in the parts per billion (ppb) level, whereas surface or wastewaters may be contaminated at the parts per million (ppm) or even percent levels.

Other factors to consider include the pumping rates that can be

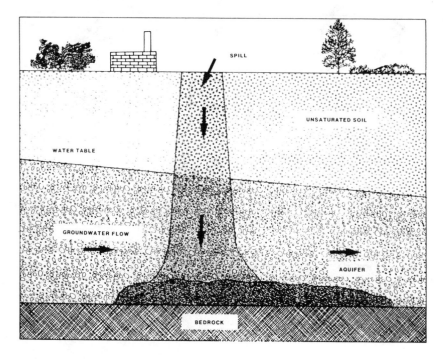

Figure 5.9.   Cross section of a nonaqueous fluid more dense than water encountering an aquifer.

achieved in a given geologic formation and the disposal of the contaminated water (whether treated or not).

An important consideration is contained in the concept of the "cone of depression." This is the response of a given aquifer system to groundwater withdrawal wells where one is attempting to remove the contaminated water. Figure 5.10 illustrates what may happen when aquifers are pumped. Because water must converge on the well from all directions and because the area through which the flow occurs decreases toward the well, the hydraulic gradient gets steeper toward the well.

Several important differences exist between the cones of depression in confined aquifers and those in unconfined aquifers. Withdrawals from an unconfined aquifer result in drainage of water from the rocks by which the water table declines as the cone of depression forms. Because the storage coefficient of an unconfined aquifer equals the specific yield of the aquifer material, the cone of depression expands very slowly. On the other hand, dewatering of the aquifer results in a decrease in trans-

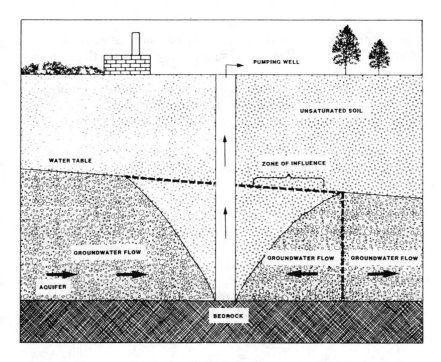

Figure 5.10.   Cone of depression in an aquifer from a pumping well.

missivity, which causes, in turn, an increase in drawdown both in the well and in the aquifer. Under extreme withdrawal conditions and low groundwater velocities, it is possible to "pump the well dry," even if only as a temporary condition.

Withdrawals from a confined aquifer cause a drawdown in artesian pressure but do not (normally) cause a dewatering of the aquifer.

MITIGATION AND TREATMENT

Since the spread of contaminants is usually confined to a plume, only localized areas of an aquifer need to be reclaimed and restored. Cleanup of a contaminated aquifer, however, is often troublesome, time-consuming, and costly. The original source of contamination can be eliminated, but the complete restoration of the groundwater is fraught with additional problems, such as defining the site's subsurface environment, locating potential contamination sources, defining potential contaminant transport pathways, determining contaminant extent and concentration, and choosing and implementing an effective remedial process.

Although not a simple task, cleanup is possible. Various methods are being used and have proven successful in certain cases. These efforts have ranged from containment to destruction of the contaminants, either in their original position in the aquifer or by withdrawing the groundwater. Examples of these remedial methods include installing pumping wells to remove contaminated water, building trenches to arrest only the contaminated flow, and stimulating biodegradation of groundwater contaminants.[5]

In many circumstances, the most reasonable and economic remedial approach is to treat the water to attain the necessary quality for a specific use. This treated water may then be used or returned to the aquifer. Certainly, combinations of barriers and treatment methods should be considered. Source control, physical control, and treatment methods all will have their part in mitigating groundwater contamination problems. Legal implications may also dictate strategies that may be utilized.

Source Control Methodologies

Source control methods include techniques to control the water source to reduce or eliminate the compounds once the groundwater has been found to be contaminated. Three methods frequently involve low capital costs and require relatively simple implementation and operation: (1) elimination of the compound at the source; (2) location of a new water supply source; and (3) blending.

The first requires identification and elimination of the contaminated source (e.g., a leaking underground tank). In a simple case, the tank is removed and the well is pumped to remove the waste. However, there are limitations which can prohibit use of this method: the source may not be easily identified and/or the aquifer size and degree of infiltration may be too great.

With the second method, location of a new water supply, an affected drinking water well is abandoned and the water supply is obtained by either developing a new well in an unaffected aquifer, utilizing an unaffected surface water supply source, or purchasing water from another area. Limitations can include prohibitive costs for new well development, the potential for contaminant migration to the new supply, or lack of sufficient water in neighboring wells to supply the affected area.

The last source control method involves blending of the water from several wells to dilute the compound to acceptable levels. Several disadvantages apply: the groundwater system may not allow sufficient blending; the organic compound concentration may be too high to achieve an acceptable dilution level; and consumers may not accept this method, since the contaminants are not removed from the system.

Source control methodologies are best utilized for small groundwater systems but can ultimately be more costly than treatment or are often unacceptable for political reasons.

Physical Control Methodologies

While there are several types of physical control methods currently used, this discussion will address abatement rather than preventative measures. There are two primary types: well systems and interceptor trenches.

Well systems manipulate the subsurface hydraulic gradient through injection or withdrawal of water. They are designed to control the movement of the groundwater directly and of the subsurface pollutants indirectly. All require installation of wells at selected sites. It is necessary to first conduct a hydrogeologic study to determine the characteristics of the contaminant plume (width, length, depth, and general shape), the hydraulic gradient across the plume, and the hydrogeologic characteristics of the aquifer. There are three major types: well point, deep well, and pressure ridge systems. Product recovery systems should also be considered. The primary obstacle is that the recovery system must be located near enough to the downgradient plume to reverse the hydraulic gradient at the boundary.

Both well point and deep well systems withdraw water. The former utilizes closely spaced, shallow wells, each connected to a main pipe (header) which is connected to a centrally located suction lift pump. The well point system is used only for shallow water table aquifers.

Deep well systems are similar to the above but they are used for greater depths and are most often pumped individually. Both should be designed so that the radius of influence of the system completely intercepts the contaminant plume.

The principle of pressure ridge systems is the inverse of that of the above two systems; water is injected to form an upconing of the water table which acts as a barrier to groundwater flow. The best application of this method is in coastal areas to prevent saltwater intrusion. The basic design principle of a recharge well is seen in Figure 5.11.

Product recovery systems most often involve hydrocarbon products (e.g., gasoline) since they tend to float on the upper surface of an aquifer. There is a strong economic incentive for cleanup as the product recovered can be used. Compounds which are not water-soluble and have a lower density than water are good candidates for this type of recovery.

Well systems in general are the most utilized methods of groundwater pollution removal. Some disadvantages of these approaches are the high operation and maintenance costs as well as the fact that clean water is removed along with contaminated and that surface treatment is often required. But there are advantages, too. Well systems can be an efficient and effective means of assuring groundwater pollution control. (However, determination of the plume's [or plumes'] characteristics can be unattainable, and the geology of the aquifer can be complex; therefore,

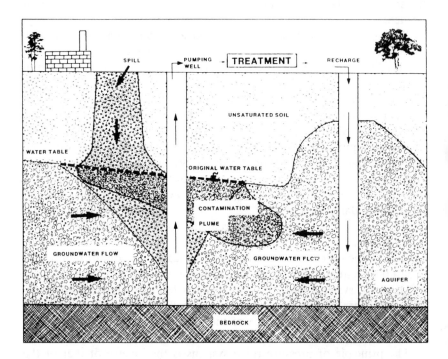

Figure 5.11.   Effect of a recharge well on groundwater in flow patterns.

removal of the total plume(s) would be impossible.) Well systems can often be readily installed, and existing monitoring wells can sometimes be utilized.

Interceptor systems are surface or subsurface trenches which can be installed on either side of a contamination site or around the site perimeter. They act like an infinite line of extraction wells; they cause a continuous zone of depression which runs the length of the drainage trench. They must be constructed at least three or four feet below the water table; they can be either active or passive systems.

Active systems will have intermediately positioned vertical removal wells or a perforated, horizontal removal pipe and are usually backfilled with coarse material to maintain trench integrity; passive systems are usually left open with the installation of a skimming pump for removal of the pollutant only. Pumping (active system) and skimming (passive system) must be continuous or the contaminants may seep into the trench walls and continue downgradient.

Passive systems are usually utilized only for those materials which are less dense than water (e.g., oils and hydrocarbons). Figure 5.12 reflects the basic principle of the interceptor systems. Capital costs can be relatively low, construction methods are fairly simple, and the system is reliable if monitoring is continuous. On the other hand, trench construction may not be feasible for some sites, it is not suitable for soils with low permeability, and operating and maintenance costs can be quite high. Also, the contaminated aquifer must be fairly shallow. As with well

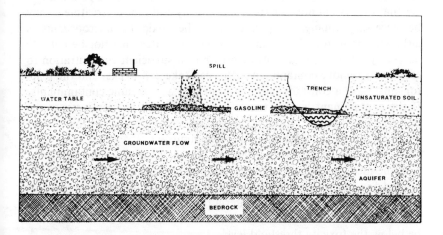

Figure 5.12.   Floating product removal by an intercepting trench.

point systems, an initial hydrogeologic study is necessary; the findings may indicate that the geology of the site is too complex and/or the plume characterization is not possible so that the use of this method may be severely limited or not applicable.

TREATMENT TECHNOLOGIES

Three types of technologies are widely used for treatment of contamination in groundwater: biological treatment, physical treatment (aeration and adsorption), and chemical treatment (oxidation and precipitation/filtration). The three types can not be utilized similarly for removal of all contaminants; removal efficiencies differ with contaminant type and concentration. All can be expensive and adsorption typically is the most costly for both capital and operating and maintenance costs.

Biological Treatment Methods

Biodegradation is the use of microorganisms for the breakdown of organic compounds and can be the lowest-cost treatment per pound of "organic" removed. Bacteria use dissolved organic compounds as sources of carbon and energy to produce more biomass. The primary limitations are that the system must be run continuously and that pretreatment may be necessary (many compounds may be toxic to microorganisms although, as seen earlier, removal of groundwater may not be entirely necessary). The bacteria must be grown to a sufficient concentration to effectively remove contaminants (this can take 2–8 weeks) and the reactor must have a constant source of nutrients and organic contaminants. If the contaminant level is too low, no effect will be seen because the bacteria may not get "enough energy out" compared to the amount of energy required to metabolize the compound.

Bacteria can be grown in either aerobic or anaerobic environments. Enzyme-mediated reactions drive the systems. Anaerobic systems are limited for use with groundwater contamination with the exception of landfill leachates. The general criteria for growth for both types are: pH = 6–9; temperature = 45–105°F; and all toxic organics and metals must be below the toxicity threshold level.

In situ biological treatment will not be discussed here since it is generally limited in application. Some limited success has been achieved, however, with peroxide injection systems for nonhalogenated organics.

There are two categories of above-ground systems: suspended growth and fixed film. Suspended growth systems are activated sludge processes. The basic system is a large basin into which the contaminated water and air or oxygen are introduced. Gravity settling is used to separate the organic material from the liquid stream. Fixed film systems have the potential to be less costly than suspended growth systems, as there is no requirement for aeration equipment. Additionally, they may be operated anaerobically, which may be advantageous to specific contaminants.

There are two types of fixed film systems: biological towers (a modification of trickling filters) and rotating biological contactors (RBC). With the former, a medium is stacked in towers approximately 16–20 feet high; a slime layer containing the microorganisms forms on the medium. The contaminated water is sprayed across the top and the contaminants are removed as the waste flows over the slime. RBC systems consist of a series of rotating discs, connected by a shaft, set in a basin or trough. The bacteria attach to the discs and metabolize the organics present in the water. About 40% of the disc surface area is under water at all times; this allows the organisms (slime layer) to have access to both air and water.

Compounds which may be degraded include acetone, benzene, methyl ethyl ketone, vinyl chloride, phenol, and methylene chloride. Some data indicate that many of the organic priority pollutants are readily degraded by biological methods. Others active in the field state that biodegradation is the only process that involves total destruction of hydrocarbons.

Biological treatment methods should be utilized when feasible and can be inexpensive when applied correctly. However, biodegradation is not applicable to all organic compounds and the water must be treated afterward to remove all bacteria (if the water is to be used for human or animal consumption).

Consideration should also be given to whether natural processes may take care of a problem given enough time. An example of such a natural cleanup process occurred near Forest Lawn Memorial Park in Glendale, California, where in 1968 a large quantity of gasoline was discovered to have leaked from a buried pipeline into the groundwater reservoir. Estimates indicated that up to 250,000 gallons of gasoline had entered the subsurface. Numerous wells were drilled and about 50,000 gallons of free gasoline were removed from September 1968 through August 1971. The remaining gasoline was not extractable by conventional methods. However, very high concentrations of naturally occurring bacteria were observed in samples from the wells. Bacterial concentrations were directly proportional to the degree of contamination, with up to 50,000 bacteria/ml found in samples containing traces of free gasoline. These bacteria, primarily certain species of genus Pseudomonas and genus Arthrobacter, were known to metabolize some petroleum products.

Laboratory tests performed by the Western Oil and Gas Association (WOGA) showed that these bacteria exhibit rapid growth rates in the presence of gasoline, oxygen (free or dissolved), and traces of naturally occurring nutrient salts. Under ideal laboratory conditions, 2 grams of gasoline yielded 1 gram of bacterial cells, with the remaining carbon converted to carbon dioxide gas. Gasoline degradation was rapid in the vadose zone and much slower in the saturated zone. The bacteria adhered strongly to the soil grains during flushing, suggesting that under natural conditions they would not be distributed by fluctuating groundwater levels. In fact, a fluctuating groundwater table would allow aeration of the entire contaminated zone, thereby accelerating the decomposition of the gasoline.

A series of progress reports to the Los Angeles Regional Water Quality Control Board (LARWQCB) from March 1971 through September 1976 showed decreasing concentrations of gasoline in the wells. A lack of detectable levels of gasoline prompted the LARWQCB to terminate the monitoring program on December 31, 1976.[6,7]

Now, whether it is politically acceptable to leave cleanup of hazardous materials spills to "nature" is another story.

Physical Treatment Methods

Aeration

Aeration refers to two distinct but analogous mass transfer processes: gas adsorption and air stripping. Gas adsorption is the process of adding gases (e.g., oxygen, carbon dioxide) to water. Air stripping is an effective process for removal of many halogenated organic compounds from contaminated water; it is a process in which a substance in solution in water is transferred to a solution in a gas (usually air).

The rate of mass transfer depends on several factors according to the following equation:

$$N_i = K_i L (C_i - P_i/H_i)$$

where N_i = mass flux of the compound i (mol/m²/hr);
K_iL = overall mass transfer coefficient (m/hr);
C_i = solute concentration of the compound i (mol/m³);
P_i = partial pressure of the compound i in the air (atm);
H_i = Henry's Law constant (atm-m³/mol).
[H_i = $P_{i,s}$ (vapor pressure of i (atm))/$C_{i,s}$ (solubility in water (mol/m³))]

The Henry's Law constant is strongly influenced by temperature and increases approximately threefold for every 10°C rise in temperature for most volatile hydrocarbons. It has been demonstrated that compounds with larger values of H_i are in general more easily "stripped." Those with H_i greater than about 1/1000 atm-m³/mol are good candidates for removal by air stripping. A high H_i indicates equilibrium favoring the gaseous phase.

There are two general categories of air strippers: diffused aerators (bubbles of air are injected into the water by means of submerged diffusers or porous plates) and waterfall aerators. Of the latter there are four basic types: cascades, multiple tray, spray basins, and packed towers. The first three have been used extensively for carbon dioxide removal in water treatment but do not provide maximum efficiency for volatile organic compound removal. Therefore, the following discussion will be limited to the packed tower column technique (specifically, countercurrent packed towers).

The terms packed tower and air stripper have often been used interchangeably as this method has been found to be the most acceptable in remedial groundwater contamination work. Another packed tower type, the cross-flow tower, is not usually used for air stripping of contaminated groundwater and is therefore not discussed here.

Figure 5.13 represents a schematic of the countercurrent packed tower design. Water is introduced at the top of the tower and flows down through the packing; the air stream flows upward picking up the volatile compounds. The air exits at the tower top with the volatiles and the water is collected at the bottom and pumped to its final destination.

The three basic design criteria for sizing of a packed column are tower diameter, tower height (packing height) and air-to-water ratio. In addition, temperature controls the stripping process and can be changed with preheaters or steam injection. The basic design begins with the mass transfer coefficient of the contaminant in the groundwater.

The primary limitation for air stripping is the release of volatile contaminants to the atmosphere. Potential problems of corrosion and

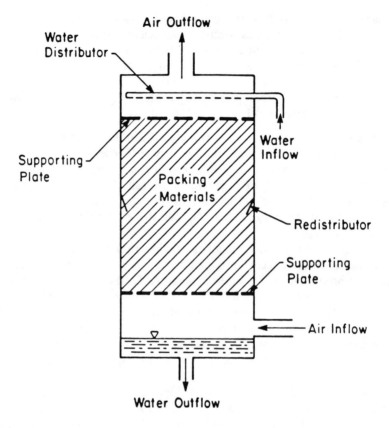

Figure 5.13.   Basic portions of a countercurrent packed column air stripper.

biofouling of the media do exist. However, these problems can be minimized with the use of corrosion inhibitors such as sodium hexametaphosphate and disinfectants such as chlorine or sodium hypochlorite.

Adsorption

Adsorption is the process of adhesion of molecules of solute to the surface of an adsorbent due to physical and/or chemical processes. Adsorptivity increases with decreasing solubility. The pH effects are compound-specific; e.g., organic acids adsorb better with acid pH and amino compounds favor alkaline pH. Aromatics and halogenated compounds adsorb better than aliphatics. Also, adsorption capacity decreases with increasing temperature, although the rate may increase. Likewise, the character of the adsorption surface has a major effect on the adsorptive capacity and rate.

Three adsorption techniques which have been considered for removal of organics from groundwater are granular activated carbon (GAC), powdered activated carbon (PAC), and synthetic resins. Activated carbon is classified as PAC or GAC according to grain size. Preliminary data indicate that very large amounts of PAC are necessary to achieve satisfactory removal efficiencies and that PAC may be effective in removing higher weight molecular compounds; but lower weight compounds (often found in groundwaters) are not satisfactorily removed.

Preliminary studies indicate that synthetic resins may be as effective as or more effective than GAC for removal of organics. They are much higher in cost per pound than GAC ($10/lb of resin to $0.65/lb of GAC) but can be regenerated in situ with low-temperature steam. Although the costs appear to be prohibitive, regeneration cost, smaller contact tanks, shorter contact times, and higher adsorption capacity per pound may make the resins a more cost-effective alternative for selected situations.

GAC is widely used for removal of organics in groundwater. The adsorption process is greatly enhanced by the large surface area/unit mass of GAC. Typical GAC surface areas range from 950 m^2/g up to 1400 m^2/g. The three steps in the adsorption mechanism are diffusion of the molecules through the liquid phase to the particle of carbon; diffusion of the molecules to the binding site; and adsorption of the molecule to the surface. The rate of each step, and thus the time required for the process, is determined by the molecule's characteristics.

Two evaluation procedures must be conducted to assess the feasibility of the use of activated carbon adsorption: a liquid-phase adsorption isotherm test and a dynamic column study. Briefly, the former is a batch test designed to demonstrate the degree to which a particular adsorbate is adsorbed on the adsorbent. A straight-line plot can be obtained for adsorbables using the empirical Freundlich equation. (The presence of a nonadsorbable component results in curvature of the line.) Dynamic testing is necessary to determine the optimum contact time and mass transfer zone (dependent upon the rate of adsorption).

Batch, column, or fluidized bed operations can be utilized with GAC; the usual contacting systems are fixed bed or countercurrent moving beds. The first may utilize downflow or upflow of water; moving beds employ upflow of water and downflow of carbon (the carbon can be moved by gravity).

As discussed previously, capital and operating and maintenance (O&M) costs of carbon adsorption systems are very high. Factors that contribute to prohibitive O&M costs include the need for frequent regeneration or replacement of the carbon due to fouling (in addition, this may require hazardous waste manifesting and hauling) and/or loss of adsorptive capacity (available binding sites are used up over time).

Chemical Treatment Methods

Chemical treatment methods for inorganic contamination are generally those applied to treatment of waste and wastewater (see Chapter 9), only scaled down to the lower concentrations found in groundwater.

These methods range from chemical oxidation (cyanide contamination) through neutralization (pH adjustment with acid or alkali) to precipitation and filtration of solids.

REFERENCES

1. Canter, L. W., and R. C. Knox. *Ground Water Pollution Control* (Chelsea, MI: Lewis Publishers, Inc., 1985).
2. Driscoll, F. G. *Groundwater and Wells,* 2nd ed. (St. Paul, MN: Johnson Division of UOP, Inc., 1986).
3. "Protection of Public Water Supplies from Groundwater Contamination," U.S. EPA Technology Transfer Publication 625/4–85/016 (1985), p. 73.
4. Kazman, R. G. *Modern Hydrology* (New York: Harper and Row, Publishers, Inc., 1965), p. 150ff.
5. Nyer, E. K. *Groundwater Treatment Technology* (New York: Van Nostrand Reinhold Company, 1985).
6. Williams, D. E., and D. G. Wilder. "Gasoline Pollution of a Groundwater Reservoir: A Case History," *Ground Water* 9(6) (1971).
7. Mackowski, M., Los Angeles Department of Water and Power. Personal communication (April 26, 1988).

6   TRANSPORTATION OF HAZARDOUS MATERIALS

INTRODUCTION

Transportation of hazardous materials is a prime source of risks of exposure to toxic contaminants. Between the generator or user of a hazardous material and the final end of that material is the transporter. Accordingly, a key link in the overall structure of hazardous materials management is safe transportation.

Under federal law, a complete set of regulations is administered by the Department of Transportation (DOT) under 49 CFR Parts 100 through 177.[1] Since transportation of hazardous materials provides a much greater potential for exposure, and since the risk is potentially more disastrous than a disposal activity, all hazardous substances are included under the DOT regulatory scheme. In distinction, EPA regulations, which apply to only hazardous wastes, do not apply for certain hazardous materials.

Nationwide DOT standards were adopted because an incident such as a spill or a fire from a transportation accident can pose an immediate threat anywhere in a transportation corridor. These federal regulations allow for tracking of hazardous materials from generation point to destination and provide some means for preventing midnight dumping of waste materials. They were incorporated in the Hazardous Materials Transportation Act.[2]

While transportation of all hazardous materials comes under DOT regulation, the EPA requires transporters to meet EPA's rules regarding hazardous wastes. These latter required activities include:

83

1. manifesting of hazardous wastes
2. labeling of hazardous wastes
3. labeling of waste materials
4. recordkeeping at the source and disposal points of hazardous materials
5. allowing only permitted facilities to be recipients of the transported hazardous wastes

The second major distinction in the transportation laws is that a federal permit is not required of transporters. This distinction is reasonable since transporters or haulers pass through many jurisdictions.

LAWS INVOLVED IN HAZARDOUS MATERIALS TRANSPORTATION

The major sections of 49 CFR (Transportation of Hazardous Materials)[3] are listed in Table 6.1. Part 171 of 49 CFR deals with general information and provides definitions of terms used throughout the regulations. Key sections are 171.8 (legal definitions), 171.10 (flammable materials aboard vessels), 171.14 (specifications for markings), and 171.15, which lists the information required in the event of a hazardous materials incident.

Section 171.17 deals with the specific notice to be given in the event of a discharge of a hazardous substance by a transporter. This discharge must be reported to the U.S. Coast Guard National Response Center immediately upon occurrence. The toll-free phone number is (800) 424–8802. The information required must be that listed in Section 171.159 as well as the name of the shipper, the quantity of material spilled or discharged, the person in charge of the vessel or transporting vehicle, an estimate of the quantity of hazardous substances removed from the scene, the manner of disposition, and any unremoved hazardous substance still left at the scene of the incident.

A summary of the hazardous materials definitions from Part 171 of the regulations (provided by the DOT) is given in Table 6.2.

Part 172 contains the bulk of the information necessary for a person involved in transportation of hazardous materials. The sections in part 172 have been divided into subparts labeled B through F dealing with the specifics on tables, shipping papers, marking, labeling, and placarding.

Subpart B (Section 172.101–102) covers 200 pages and consists primarily of the Hazardous Materials Table and explanations of the sym-

Table 6.1 Transportation of Hazardous Materials: 49 CFR Titles

Part	Section		Title
171			General information, regulations, and definitions (see Table 6.2)
172			Hazardous materials tables and hazardous materials communications regulations
	Subpart B	(.101–.102)	Tables, descriptions, names, classes, labels, and packaging
	Subpart C	(.200–.205)	Shipping papers
	Subpart D	(.300–.338)	Marking
	Subpart E	(.400–.450)	Labeling
	Subpart F	(.500–.556)	Placarding
173			Shippers requirements (specific packaging by material)
174			Carriages by rail
175			Carriages by aircraft
176			Carriages by vessel
177			Carriages by public highway

bols used in that table. Also included is a listing of the UN ID numbers for identifying hazardous materials in a cross-reference to their proper shipping name used in the hazardous materials tables. A copy of one page from that table is seen in Figure 6.1.

In Figure 6.1 we note that the substances are classified by their chemical name, chemical category, labeling, packaging, and usage, as well as by additional information regarding shipments over water. Each of these columns becomes important when filling out the manifest and labeling the containers for transport of hazardous materials.

Marginal notes such as +, E, A, and W determine when the material is subject to regulation or some other legal requirement. For instance, the "+" fixes the proper DOT shipping name which must be on all paperwork associated with the transport of that material. "E" means that material is subject to regulation by DOT regardless of the mode of transport, whereas "A" and "W" refer to DOT regulations only if transported by air or water. For instance, cotton is subject to regulation by DOT only if transported by water. No label is required; nevertheless, its hazard

86   HAZARDOUS MATERIALS MANAGEMENT

Table 6.2 Hazardous Materials Definitions – DOT

Hazardous material – A substance or material, including a hazardous substance, which has been determined by the Secretary of Transportation to be capable of posing an unreasonable risk to health, safety, and property when transported in commerce, and which has been so designated.

Hazardous substance – A material and its mixtures or solutions that is listed in the Appendix to Section 172.101, and is in a quantity, in one package, which equals or exceeds the reportable quantity (RQ). (Does not apply to fuels or lubricants.)

Hazardous waste – Any material that is (1) subject to the hazardous waste manifest requirement of the Environmental Protection Agency specified in the CFR, Title 40, Part 262; or (2) would be subject to these requirements.

Multiple hazards – A material meeting the definition of more than one hazard class is classed according to its position in the lists in Section 173.2(a) and (b) (HMTA).

Reportable quantity – The quantity (of hazardous substance) specified in the Hazardous Materials Table (Sec. 172.101). Reportable Quantity is identified by the letters "RQ."

Explosive – Any chemical compound, mixture, or device designed to function by explosion that is substantially instantaneous with the release of gas and heat.

Source: "Hazardous Materials Definitions," U.S. Department of Transportation, Research and Special Programs Administration, Office of Hazardous Materials Transportation, Transportation Resources Branch (October 1987).

class is "other regulated materials" (ORM)-C (the reason being that cotton aboard a ship could present a fire hazard under the right conditions).

The "RQ" refers to a "reportable quantity" and designates the weight (in pounds or kilograms) of the material which qualifies the substance to be a reportable quantity for purposes of shipment and regulation.

Subpart C refers to shipping papers for all modes of transportation and includes by reference the hazardous waste manifest required under RCRA regulation 40 CFR Part 262.

Subpart D (Section 172.300–338) deals with very specific information regarding marking of packages, freight containers, and transport vehi-

Hazardous Materials Table—Continued

(1) +/E/A/W	(2) Hazardous materials descriptions and proper shipping names	(3) Hazard class	(3A) Identification number	(4) Label(s) required (if not excepted)	(5) Packaging (a) Exceptions	(5) Packaging (b) Specific requirements	(6) Max net quantity in one package (a) Passenger carrying aircraft or railcar	(6) (b) Cargo aircraft only	(a) Cargo vessel	(b) Passenger vessel	(7) Water shipments (c) Other requirements
EA	Copper based pesticide, solid, n.o.s. (compounds and preparations),	Poison B	UN2775	Poison	173.384	173.385	50 pounds	200 pounds	1,2	1,2	
	Copper chloride (RQ-10/4.54),	ORM-B	UN2802	None	173.505	173.800	25 pounds	100 pounds	1,2	1,2	
	Copper cyanide	Poison B	UN1587	Poison	173.370	173.370	25 pounds	200 pounds	1,2	1,2	Stow away from acids
	Copper tetramine nitrate	Forbidden									
W	Copra	ORM-C	UN1363	None	173.505	173.960		200 pounds	1,2	1,2	Segregation same as for flammable solids. Separate from flammable gases or liquids, oxidizing materials, or organic peroxides
	Copra pellets. See Coconut meal pellets										
	Cordeau detonant fuse	Class C explosive		Explosive C	None	173.104	50 pounds	300 pounds	1,2	4	
	Corrosive liquid, n.o.s.	Corrosive material	UN1760	Corrosive	173.244	173.245 173.245a	1 quart	1 quart	1	4	For material that meets only the corrosion to skin criteria of 49 CFR 173.240(a)(1), under deck' storage is also authorized if the description includes the additional entry specified by 172.203(i)(2)
	Corrosive liquid, poisonous, n.o.s.	Corrosive material	UN2922	Corrosive and Poison	173.244	173.245	1 quart	1 quart	1	4	
	Corrosive solid, n.o.s.	Corrosive material	UN1759	Corrosive	173.244	173.245b	25 pounds	100 pounds	1	4	For material that meets only the corrosion to skin criteria of 49 CFR 173.240(a)(1), under deck' storage is also authorized if the description includes the additional entry specified by 172.203(i)(2)
	Cosmetics, liquid, n.o.s.	Corrosive material	NA1760	Corrosive	173.244	173.245	1 quart	1 quart	1,2	1,2	
	Cosmetics, n.o.s.	Combustible liquid	NA1993	None	173.118a	None	No limit	No limit	1,2	1,2	
	Cosmetics, n.o.s.	Flammable liquid	NA1993	Flammable liquid	173.118	173.119	1 quart	10 gallons	1,2	1	
	Cosmetics, n.o.s.	Flammable solid	NA1325	Flammable solid	173.153	173.154	25 pounds	100 pounds	1,2	1,2	
	Cosmetics, n.o.s.	Oxidizer	NA1479	Oxidizer	173.153	173.154	25 pounds	100 pounds	1,2	1,2	
	Cosmetics, solid, n.o.s.	Corrosive material	NA1759	Corrosive	173.244	173.245b	25 pounds	100 pounds	1,2	1,2	
W	Cotton	ORM-C		None	173.505	173.965			1,2	1,2	Keep dry
E	Coumaphos (RQ-10/4.54)	Poison B	NA2783	Poison	173.364	173.365	50 pounds	200 pounds	1,2	1,2	Segregation same as for flammable solids. See 176.900 to 176.904
E	Coumaphos mixture, liquid (RQ-10/4.54)	Poison B	NA2783	Poison	173.345	173.346	1 pint	1 quart	1,2	1,2	
E	Cresol (RQ-1000/454)	Corrosive material	UN2076	Corrosive	173.245	173.245	1 quart	10 gallons	1,2	1,2	

Figure 6.1. A page from the Hazardous Materials Table.

cles containing hazardous materials. These requirements range from general marking requirements to export shipments by water and the requirements for color and contrast for packages containing ORM materials or hazardous substances. Requirements for radioactive materials, cargo tanks, and general ID numbers for combustible and other materials are also included.

Subpart E deals with labeling requirements and gives specific instructions regarding when they are necessary, their size and legibility, and, in particular, the color specifications.

Subpart F (Section 172.500–558) deals exclusively with placarding of vehicles that transport hazardous materials and gives detailed instructions regarding the size, dimensions, and coloring of all of the types of placards required.

The other major parts of 49 CFR deal with specific information for transport of hazardous materials, classified by type of carriage. Part 173 deals with general shipment requirements, 174 with carriage by railroad, 175 with carriage by aircraft, 176 with carriage by oceangoing vessels, and 177 with carriage by public highway.

Motor Carrier Safety Act

Transportation vehicles are regulated by the DOT under the Motor Carrier Safety Act[4] whenever a transporter is carrying any hazardous material.

Vehicles are subject to rigid safety standards prior to operating on a public highway.[5] Major components of required vehicle checks are the braking and steering systems, electrical systems, tires, lights, mirrors, and door locks, plus any valves and piping for tankage vehicles. Dual safety precautions are required both in loading and unloading of the vehicle, as well as use of personal protective gear. Whether carrying bulk liquids or solids or drummed shipments, safety must be uppermost in the mind of the transporter.

The Motor Carrier Safety Act also requires safety training and driver training for those persons involved in highway transport of hazardous materials, and identification of the materials being carried on that vehicle.

PRACTICAL ASPECTS OF HAZARDOUS MATERIALS TRANSPORTATION

A key to properly filling out all of the forms and materials required for transport is a thorough understanding of the above tables and how they relate to each other. Also important is the cross-linkage of the RCRA hazardous waste regulations regarding manifesting and recordkeeping. For a transporter to legally perform the job, an EPA ID number must be obtained. This is administered by EPA using EPA Form number 8700-12.

The Waste Characterization Sheet

Every disposal site requires a waste characterization sheet prior to shipment of hazardous wastes to the facility. It should be remembered that federal regulations require that waste materials be sent only to a licensed facility. Without prior approval and scheduled delivery of waste materials, licensed TSDFs will not accept the shipment.

The Hazardous Waste Manifest

The hazardous waste manifest is a legal document. Failure to properly complete a hazardous waste manifest can result in fines and other regulatory action. The persons involved in handling any manifested shipment of hazardous waste are the generator, the transporter, and the owner or operator of the TSDF that receives the waste shipment. (For hazardous materials, only shipping papers are required.) A copy of the hazardous waste manifest is shown in Figure 6.2.

In general, the generator of hazardous waste is responsible for supplying most of the information contained on the manifest. This block of information characterizes the waste and identifies all parties involved in the generation, transport, treatment, and disposal of the waste. It is a key element in the complex paper trail that must accompany a waste shipment from generation to final disposal. When a transporter takes possession of the waste, he or she is required to sign and date the manifest. Though the manifest serves as a system of checks and balances to all parties in the industry, it is especially important to the transporter. The transporter's job depends on an accurate manifest.

When the waste material is eventually received at a treatment or dis-

Please print or type *(Form designed for use on elite (12-pitch) typewriter.)*

Form Approved. OMB No. 2050-0039. Expires 9-30-88

UNIFORM HAZARDOUS WASTE MANIFEST

1. Generator's US EPA ID No. Manifest Document No.
2. Page 1 of

Information in the shaded areas is not required by Federal law.

GENERATOR

3. Generator's Name and Mailing Address
4. Generator's Phone ( )
5. Transporter 1 Company Name
6. US EPA ID Number
7. Transporter 2 Company Name
8. US EPA ID Number
9. Designated Facility Name and Site Address
10. US EPA ID Number

A. State Manifest Document Number

B. State Generator's ID

C. State Transporter's ID

D. Transporter's Phone

E. State Transporter's ID

F. Transporter's Phone

G. State Facility's ID

H. Facility's Phone

11. US DOT Description *(Including Proper Shipping Name, Hazard Class, and ID Number)*	12. Containers		13. Total Quantity	14. Unit Wt/Vol	I. Waste No.
	No.	Type			
a.					
b.					
c.					
d.					

J. Additional Descriptions for Materials Listed Above

K. Handling Codes for Wastes Listed Above

15. Special Handling Instructions and Additional Information

16. **GENERATOR'S CERTIFICATION:** I hereby declare that the contents of this consignment are fully and accurately described above by proper shipping name and are classified, packed, marked, and labeled, and are in all respects in proper condition for transport by highway according to applicable international and national government regulations.

If I am a large quantity generator, I certify that I have a program in place to reduce the volume and toxicity of waste generated to the degree I have determined to be economically practicable and that I have selected the practicable method of treatment, storage, or disposal currently available to me which minimizes the present and future threat to human health and the environment; **OR**, if I am a small quantity generator, I have made a good faith effort to minimize my waste generation and select the best waste management method that is available to me and that I can afford.

Printed/Typed Name | Signature X | *Month Day Year*

TRANSPORTER

17. Transporter 1 Acknowledgement of Receipt of Materials

Printed/Typed Name | Signature | *Month Day Year*

18. Transporter 2 Acknowledgement of Receipt of Materials

Printed/Typed Name | Signature | *Month Day Year*

FACILITY

19. Discrepancy Indication Space

20. Facility Owner or Operator Certification of receipt of hazardous materials covered by this manifest except as noted in Item 19.

Printed/Typed Name | Signature | *Month Day Year*

EPA Form 8700-22 (Rev. 9-86) Previous editions are obsolete

Figure 6.2. A hazardous waste manifest.

posal facility, a facility representative is required to sign and date the manifest. Thus, the manifest becomes a complete record of who has handled the waste, what it is, and where it is going, and it is proof of acceptance by the treatment or disposal facility.

All information placed on a manifest must be legibly printed or typed. All required information sections must be complete and correct. However, when corrections are made on a manifest, the person making that correction must initial and date the space corrected. Manifests which have numerous strikeovers or corrections should be completely redone.

When five or more waste streams are being combined into one shipment, a uniform hazardous waste manifest continuation sheet must be used. There is no limit on the number of pages that may accompany any single shipment of hazardous waste. The manifest must accompany the waste even if it is not an EPA waste, i.e., even if it is a state hazardous waste only. The only exception would be if a specific exemption exists to the manifesting requirement. It should be noted that a transporter of hazardous waste may not accept an incorrect manifest, but may correct the manifest. The transporter's initials and the date must be placed on the document.

All manifest entries must be made using complete DOT descriptions and they must be in their correct sequential form: waste name, classification by hazard, and UN ID number. Other abbreviations may not be used unless specifically authorized. Some of the approved abbreviations include:

- NOS = not otherwise specified
- NOIBN = not otherwise indexed by name
- LTD QTY = limited quantity (refers to amounts limited for specific packaging purposes)
- RQ = reportable quantity; to be shown if the amount of material being shipped exceeds a specified reportable quantity under DOT regulations

When a reportable quantity designation is required, the letters RQ must appear on the waste label as well as on the manifest. Placement of the letters RQ *must* be either before or after the proper DOT description. The RQ is based on the amount of reportable quantity material present in an individual container: either a single drum, or the amount in the individual vehicle if the waste is not containerized (a bulk shipment). The RQ value for each compound is listed in the DOT hazardous materials table. It depends upon the type of material and its chemical characteristics. Some waste materials have an RQ of 5000 pounds; if 5000 pounds of such a material were to be shipped in individual 55-gallon drums, the RQ value would not be exceeded in any single drum. In that particular case, an RQ would not be required, since individual drums do not generally carry that much mass.

Care must be taken in filling out the manifest; something as simple as a spelling error can be a violation of federal law, since it is not a proper reproduction of the basic DOT description. Each individual item must be correctly completed in full on the manifest.

Also, by signing the document, the generator certifies that he or she has a program in place to reduce the volume and toxicity of waste generated at the facility. While no penalties have been instituted to date, it is a certification that a generator is making ongoing efforts at his own facility to reduce the amount of waste being shipped.

Packaging and Labeling

Correct and accurate packaging and labeling are specified in the hazardous materials table and are required by federal law. A vital aspect of the hazardous materials management process is good packaging. It must be containerized and labeled in a way that provides for safe storage, transport, and eventual disposal. Hazardous waste labeling requirements for packaging are very specific. Knowing these elements contributes to a good hazardous materials management system.

In general, most hazardous materials are transported either in bulk solid deliveries, bulk tanker deliveries (by either rail or highway), or the familiar 55-gallon steel drum. For containerized solids, the DOT stamp "17H" appears upon those 55-gallon drums approved for solid waste materials. The 17H drums (17C are open-headed) should be inspected for integrity prior to usage. Corrosion, dents, obvious splits, or cracks indicate that the drums should be rejected prior to waste actually being containerized.

If waste materials are designated as a solid, then no liquids may be present in that containerized drum. When the waste material is destined for a landfill, the law requires that the container be at least 90% full.

Bringing the total volume of material into the drum up to 90% or greater using a material such as sawdust is not allowed under federal regulations. Only specific absorptive media or reactive media that will solidify or immobilize the waste is allowed to be put into such a drum.

Only the waste material identified on the manifest may be placed in containers. The drum must be securely sealed and the outside must not be contaminated. After the drum is filled, it must be properly labeled and all containers marked.

However, only waste material regulated by the EPA requires a "hazardous waste" label. If such a label is required it should be placed on the side of the container near the top on the first day that waste material is placed in it. DOT warning labels, if required, must be placed near the hazardous waste label or near the DOT proper shipping name if the EPA hazardous waste label is not required.

For containerized liquids, typically the 55-gallon drum with a bunghole (stamped with the number 17E) is often used. Rusted, damaged, dented, or leaking containers should never be used. Again, only waste material identified on the manifest may be placed in the container. Proper labeling is required, as is securing the seal bungs on the drums prior to shipment. Land disposal of liquids is, of course, prohibited.

Bulk solids must be placed in a vehicle that provides for operational integrity and secured with security bolts, chains, or other locking mechanisms in the event the primary locking system fails. The use of synthetic liners in bulk containers such as roll-offs and dump vehicles is a good practice. No free liquids may ever be present in bulk solid shipments. As is required, only identified waste material should be in a bulk shipment, and it should be covered to prevent water intrusion during shipment by rain, melting snow, or water draining off bridges and overpasses. The bulk container must be labeled and placarded.

Bulk liquids have the additional requirements that all valves must be equipped with security caps and that some consideration must be given to the potential for solids settling out of the waste liquid space. Sufficient void space must be provided to allow for liquid expansion to preclude tank overflow while in storage or in transit. The proper UN or North American (NA) ID number must be displayed on the bulk shipment as required.

The DOT requirements, the specified format, and color for all warning labels on packages of hazardous materials are spelled out in 49 CFR 172.407ff.

There are general trends in these labels. Combustible and flammable materials generally show a flame symbol with a red background on the label. Oxidizing materials will show a flame surrounding an "O" (for oxidizer) on a yellow background. Poisons will show a skull and crossbones symbol with a white background. Pressure-generating materials, such as explosives or reactives, will either say "explosive" and show a blast pattern, or will have the words "blasting agent." Both will have an orange background color.

Certain labels show a combination of one or more symbols. Flammable solids would be marked with a white label, a flame pattern, and cross-hatched red markings.

Biomedical materials, infectious substances, and radioactive materials have their own distinctive symbols as well as a variety of different colors.

Corrosives will always show a test tube with either a hand, a brick, or some other object being eaten away on a black and white label, along with the word "corrosive."

The hazardous waste container label has a yellow background with red lettering and borders with space for specific written information. This label goes on the container or package of hazardous waste. The label must include the proper DOT shipping name, the UN ID number, the generator name and address, the manifest document number, the accumulation start date, and the EPA ID number for the type of waste.

Placarding

Placarding requirements refer specifically to the transporter or hauler of hazardous materials.

The basic purpose of placards on vehicles transporting hazardous material is to indicate to emergency response people what material is in that shipment in the event of an accident, spill, discharge, or other transportation incident.

Placarding, when required, must be on all four sides of a transport vehicle. However, placarding may or may not be required, depending upon the type of material. In the example earlier, cotton (though listed in the DOT hazardous waste table) would not present an environmental problem in the event of a highway or rail accident; therefore, a placard would not be required on that vehicle. An ID number, however, would be required on that shipment even though no other markings were required.

Placards themselves are specified by DOT with respect to size and markings. However, they may be made of either metal, vinyl, or magnetic material. It should be noted that a generator cannot make his own placard, even though he is required by law to provide them to the transporter of hazardous waste, unless it meets the specifications of

49 CFR 172.504H. Typically, a registered hauler of hazardous material will provide them as a service to the generator.

A placard is required if:

- 1000 pounds or more of material (gross weight) is being shipped, or
- any amount of material in the class A or B explosive, poison A, flammable solid, or radioactive yellow III class is being shipped

Placards are not required for ORM classes A, B, C, D, and E.

Placards must be readily visible from the direction in which they face. They must be on the front, back, and both sides of the vehicle and cannot be obscured in any way by ladders, doors, pipes, dirt, or water spray.

Also, placards may not be affixed to a vehicle which is completely empty. Placards themselves may be removed only after a tank (in the case of bulk liquid) has been cleaned, purged, and/or reloaded with a nonhazardous material.

If there are two or more hazardous materials being transported with a weight of 1000–5000 pounds each, then the placard "dangerous" may be used (except for cargo tanks). In the event that more than 5000 pounds of any one class of material is loaded at any one facility, the placard for that material must be shown on the vehicle. If, however, any of the materials is subject to a "poison-inhalation hazard" description, then the inhalation poison or poison class placard must be used.

Colored charts for both placards and labels are available from the DOT Office of Hazardous Materials Regulation. There is a strong similarity between the name, layout, and color for placards and those features on hazardous container labels. They are not the same, however, since they differ in size as well as utilization.

The UN hazard class identification number that must go on vehicles, cargo tanks, and tank car shipments of hazardous materials must be displayed along with the placard for the type of shipment carried.

REFERENCES

1. 49 CFR 171.1.
2. Public Law 93–633, Hazardous Materials Transportation Act (1975).
3. 49 CFR 171–177.
4. 49 USC 1803.

5. "Federal Motor Carrier Safety Regulations (Parts 390–399)," U.S. Department of Transportation, American Trucking Association, Inc. (March 1984).

7 WASTE CHARACTERIZATION AND ANALYTICAL METHODS

INTRODUCTION

Whether one is involved in management of hazardous materials or treatment of hazardous waste (onsite or offsite), or is potentially facing cleanup of a contaminated site, some degree of waste sampling, characterization, and analysis by chemical means is going to be required. Unfortunately, "good data is hard to find." Certainly with respect to environmental samples, a great deal of thought, planning, and preparation must be made for a good analysis and a good characterization to be performed for a waste stream, a contaminated site, or a compliance monitoring program.

EPA REFERENCE METHODS

The EPA has its own sampling and analysis methods which must be used for projects involving the EPA. The major elements contained in SW846 (Test Methods for Evaluating Solid Waste—Physical/Chemical Methods)[1] are shown in Table 7.1. In this table we see the full range of activities associated with EPA's approach to sampling and analysis of environmental samples.

Under SW846, sampling plans are the first element. A sampling plan consists of the objectives of the project, the statistical methods used in determining the number of samples, the approach used to determine sample location, and the types of sampling plans involved in this type of work.

Table 7.1 Sampling and Analysis Methods (SW846): Major Elements

1. **Sampling plans**
   Objectives
   Statistical sampling
   Types
2. **Implementation**
   Sampling equipment
   Sample containers
   Processing and storage
3. **Chain of custody**
   Labels, seals, logs
   Shipping and receipt
4. **Sampling methodology**
5. **Characteristic tests**
   Ignitability
   Corrosivity
   Reactivity
   Extraction procedure toxicity
6. **Sample preparation**
7. **Analytical methods**
8. **Quality assurance/quality control**

Implementation is the second element and includes gathering together the sampling equipment and the sampling containers, and processing and storing the environmental samples.

An important third element, particularly with respect to regulatory or potential legal action, is the chain of custody. This involves labels, seals, and log books of when and by whom certain samples were taken, how they were shipped, and how and in what condition they were received at the analytical laboratory. This chain-of-custody element has become an important part in enforcement activities and is also important in evaluating whether a sample was received intact or in a timely manner.

The methodology of how to sample is described in the next paragraphs of the EPA document and is followed by the types of tests for hazardous waste characteristics. These include tests for ignitability, corrosivity, and reactivity, as well as the EPA Extraction Procedure toxicity test.

The next element concerns the preparation of the sample prior to analysis in the laboratory. This includes receipt, storage at the laboratory, extraction or filtering of the samples, maintaining the proper temperatures and chemical states of the samples, and bringing them to the point where they can be analyzed.

The next step and major element of SW846 is the actual analytical lab method. This states specifically how the sample, once it has been prepared, is to be analyzed for a specific constituent.

The final elements in EPA's approach are quality assurance and quality control activities. These deal with how one may be confident that the sample was taken and analyzed in a manner that both is representative and truly exhibits the concentration of the contaminant of concern. It involves taking blank and spike samples, running duplicate analyses, setting up control charts, and determining percent recovery. These are used for statistical analyses of data for legal or decisionmaking activities.

It is also important to realize that SW846 is incorporated by reference into 40 CFR Parts 260–265,[2] which refer to these methods of analysis for determining which wastes come under regulation for RCRA. Over 85 organic chemicals and over a dozen inorganic chemicals are called out for specific test methods in Appendix III of SW846. Therefore, the latest version must be used.

It must be noted that for certain elements there are a variety of chemical test methods that may be used, depending upon the required level of detection and the expected range of concentration of the particular contaminant. Chromium alone has five different test methods, depending upon its oxidation state and the type of analytical method.

FUNDAMENTALS OF CHEMISTRY FOR SAMPLE ANALYSIS

Every analytical method takes advantage of some property of a particular element or molecule. Therefore, both atomic structure and molecular structure are important in determining the type of method that can be used, its advantages, and its disadvantages. It is assumed that the reader is familiar with the basics of chemistry and physics. However, a brief review will be given at this time.

There are two major classifications of compounds in the field of chemistry: inorganic and organic.

The inorganic compounds consist of metals, noble gases, and most nonmetals. The metal elements are the alkali, alkaline earths, and transition metals. These occur both naturally and in compounds synthesized by man, and can be both raw materials and waste products.

The noble gases do not participate in any normal chemical reactions and are considered inert. Nevertheless, one of the noble gases, radon, has another property, radioactivity, not related to its chemical reaction potential. This means that radon may spontaneously break apart (undergo fission) and emit various subatomic charged particles and electromagnetic radiation.

The nonmetals typically consist of oxygen, fluorine, chlorine, and sulfur, and typically either form acid gases or participate in acid reactions in water solutions.

Organic compounds consist of those elements which are constructed with carbon as the basic linking atom in a molecule and which may or may not have hydrogen present. The carbon compounds can be linear in structure and saturated (alkanes); unsaturated (alkenes); or, like benzene, cyclic (aromatic compounds).

The combination of the organic compounds above with sulfur, nitrogen, and oxygen form other broad classes of organic compounds and include alcohols, ethers, ketones, esters, aldehydes, and amino compounds.

A second class of organic compounds is those that contain fluorine, chlorine, bromine, or iodine in addition to carbon and hydrogen atoms; these are called halogenated organic compounds.

UNITS OF MEASUREMENT AND PHYSICAL AND CHEMICAL PROPERTIES

One of the key concepts involved in environmental sampling is concentration. Concentration is a ratio and is expressed as the number of molecules or the mass of material present per unit of volume or mass. Concentrations in environmental samples may be expressed in parts per million (ppm), mg/L (milligrams per liter), or $\mu g/nm^3$ (micrograms per normal cubic meter). This concept is important, because environmental analytical methods will yield a result in terms of concentration.

The chemical properties of particular elements, whether in solution or in air, are influenced by a number of things, and there are measures of such properties. One of these is pH and is a measure of the acid or base (alkaline) strength of a water solution. This is measured on a sliding scale from 0 to 14 where 7 is perfectly neutral. All numbers from 7 to 14 are considered alkaline in increasing strength, and those numbers less than 7 are acidic, with a pH of 0 being the most acidic.

Chemical oxidation and reduction potential is another measure of the ability of various elements and ions to enter into reactions with other elements, ions, or molecules.

Physical properties that are often utilized in analyzing environmental samples include (among others): vapor pressure, partition coefficient, solubility in water, and density.

MATRICES

The specific matrix in which a contaminant may exist, whether it be a water solution, an oily water foam, or a sludge, is very important in terms of how the sample is collected, how it is "worked up," and finally how it is analyzed. Another major aspect in matrix effects refers to interferences that may occur in that analysis due to the presence of similar compounds. Also, the enhancement or reduction in sensitivity of a test method because of other contamination may affect test results.

Liquids

Solutions are among the most common matrices encountered in hazardous waste work. There are two types of solutions: organic (nonwater-based) solutions and aqueous (water-based) solutions. In environmental samples, the latter are in the majority. The nature of water itself determines how different materials may be dissolved, how they may be transported, and how they may react.

Gases

Gaseous samples are the cleanest type of environmental sample since only gases are present and once taken, they have little further reactivity (apart from reactions with their containers). However, it should be remembered that gases are sensitive to pressure and temperature effects; also, they will react in the presence of light. Sunlight in particular reacts with certain gaseous constituents to form either free radicals or other oxidized or reduced species.

Solids

Solid samples are important due to the issues surrounding waste contamination of land. Since they are heterogenous, it is very difficult to get an average analysis of a solid sample. Other impacts that a solid sample matrix may have on an analysis method are possible catalytic effects, and the effect of soil particles on absorption of contaminants or other chemicals present.

The particle size may influence the ability of that sample to be either extracted or homogenized so that a representative sample may be taken.

Mixtures

Mixtures are also common in environmental work. The mixture sample is one in which it is possible to physically separate the various phases of that mixture into discrete components. Mixtures may be due to either dissolution (as alcohol in water) or phase mixing (an emulsion).

Two- and three-phase systems may also occur and influence the sample one is interested in. These may be exemplified by API separator sludge from petroleum refineries, which consists of oil, water, and sand or silt components, along with other dissolved contaminants such as salt.

Important considerations for mixtures are density and miscibility. In the example of oil and water (with the proper amount of handling), the two phases physically separate so that one may analyze each phase separately. However, many mixtures may form an emulsion. Mixtures that present particular problems in an environmental sample are sludges and pastes. These consist of very tightly bound, highly viscous samples that are extremely difficult to extract, separate, or analyze.

DETECTION LIMITS

In a true scientific sense, a zero concentration of molecules, atoms, or ions does not exist. Given enough time and money, practically every element or molecule in the universe can be found in a given environmental sample.

Amadeus Avogadro, the Italian physicist, announced the concept of molecular volumes in 1811. From this concept, scientists have calculated the number of atoms, molecules, or ions that may occur in one "gram-mole" of a substance. This is one of the foundational concepts in physical science.

One gram-mole of a substance contains 6.02×10^{23} atoms, molecules, or ions of that substance. Because of the magnitude of the number and the use of logarithmic designation (i.e., 10^{23}), one is driven to use logarithmic terminology in understanding the number of atoms or molecules of material in a sample. When we are dealing with powers of ten we speak of "orders of magnitude" and these become the controlling numbers in discussions rather than the integers.

Because environmental samples span large concentration ranges, we use terminology that separates these logarithmic numbers by groups of three orders of magnitude. The abbreviations of the five groups most commonly used in environmental work today are:

Prefix	Exponent
Milli	10^{-3}
Micro	10^{-6}
Nano	10^{-9}
Pico	10^{-12}
Femto	10^{-15}

Thus, when one is comparing a concentration of a hazardous substance at 10^{-15} grams/kg (one femtogram/kg) with 10^{23} molecules being present in a gram-mole of material, one is lead to the fact that 10^{23} is still 100 million (10^8) times larger. Thus we can see that "0" (i.e., *no* atoms or molecules) is a very hard number to find, even when one is dealing with exceedingly small concentrations (i.e, at the femtogram level).

Based on the above information, it is important to realize that detection limits are a key with environmental samples. The lowest concentration that one can report is a concentration less than the detection limit for a given chemical compound and a given analytical chemical method. It is appropriate, then, to state what the detection limit is for a particular method and compound.

Taking chromium as an example, one can say that the detection limit is 1 μg/L by the furnace AA method, 0.5 mg/L for the flame AA method, or 0.05 mg/L for the ICP method of analysis (according to SW846). If none of those methods showed the presence of chromium in a sample, the best one could say is that the concentration was less than the lowest detection limit or, in this case, less than 1 μg/L by the furnace AA method.

Definitions

EPA's SW846 gives us two definitions of detection limit for any given type of analysis. The MDL (method detection limit) is defined as "the minimum concentration of a substance that can be measured and is reported with 99% confidence that the value is above 0." The other detection limit, the EDL (estimated detection limit) is "calculated from the minimal detectable response being equal to five times the background noise." Thus, in the reference method definitions we see that there are concentration values which may exceed zero but nevertheless cannot be determined by a given analytical method.

It is interesting to note that this definition introduces statistical concepts into the analytical methodology.[3] Of these, the concepts of precision, accuracy, and mean value are important. The mean value is the average of a number of trials or analyses run on a single sample and is reported with the standard deviation. Precision refers to how close together all analyses of a given sample are and how small the standard deviation is. Accuracy refers to how close the average value is to the true value. Since the true value is virtually unknown, one is forced to deal with precision rather than accuracy in environmental samples.[4]

Lowering Detection Limits

A given method may be enhanced by concentrating the analyte (compound or element one is looking for) in a sample. In some cases, it is possible to get a "parts per billion" value from a method that utilizes "parts per million" detection limits by taking this approach. The "purge and trap" method of determining volatile organics in a water sample is a good example. In this method, the water sample is stripped of the volatile organics which are then concentrated on a collection media. The latter is extracted, and the extract is further concentrated and then analyzed. Detection limits several orders of magnitude lower can thus be realized.

REPRESENTATIVE SAMPLING

Since most environmental samples taken in the hazardous waste field are heterogeneous, it is important that representative sampling be performed.

To ensure representative sampling, a sampling plan or protocol is required which ensures that the correct number of samples will be taken at an appropriate frequency and which minimizes sample loss or degradation.

The key item in preparing a sampling plan is to determine the explicit purpose for which one is conducting the sampling. One of the biggest problems in the environmental field is trying to meet too many objectives with one particular sampling method. Typical of the purposes that would require different analytical and sampling methods would be:

- a site survey for possible contamination
- meeting a permit discharge limit
- prequalifying a waste stream for a disposal facility
- meeting an efficiency guarantee for a pollution control device

The possibilities for matrix effects and interferences should be among the first items considered. Other considerations include where and how often the samples are to be taken, how many, and even the weather expected during the sampling. Whether the sample is to be a gas, liquid, solid, paste, sludge, or some combination thereof is important and will lead to different methodologies. Multiple samples are usually taken in order to determine a statistical average.

Sample Loss or Degradation

After sampling, one must consider storage and travel time. Settling of solid materials originally suspended in liquid when the sample was taken may also occur.

It is very important to consider phase changes that may occur in addition to the number of phases found while sampling (i.e., volatile gas losses). If the hazardous constituents can potentially exist in more than one phase (i.e., a partition coefficient between gas and sample), one must consider this in the sampling plan in order to obtain a representative sample.

For instance, a sample may contain a volatile organic compound in a solid phase or in the void spaces between the solid particles. Once the sample is placed in a glass jar with the lid closed, the volatile organics will distribute themselves into the gaseous space (head space) at the top of the jar, so that when the sample is removed some of the volatiles will be lost.

Reactions of the sample with sunlight or temperature extremes, or the potential for polymerization in the container are also important considerations in the sampling plan. Freezing of water samples in glass containers without head space has been known to break the glass, resulting in sample loss.

Preservatives are often used in environmental samples but may also interfere with the analyses. When one is using preservatives, one must ensure that they are noninterfering and that they stabilize the substance at the concentration present when the sample was taken. Table 7.2 presents practical considerations when one is taking field samples of either liquids, pastes, or solids.

Types of Samples

There are four basic types of samples, categorized according to the frequency with which they are taken and the number taken:

- continuous
- sequential
- composite
- random (or grab)

The most accurate sampling methods use continuous instrumental methods of analysis, but unfortunately, there are very few instruments available apart from certain gas analyzers. A second but potentially expensive method is the sequential method whereby multiple samples are taken for each analysis. A third method, which is a modification of the sequential sampling technique, is to form a composite or average sample by combining a number of individual samples. This is cheaper but one may get some "surprises," since high values may get lost by dilution in a composite sample. The fourth category is the purely random, or grab, sample. These samples are generally of the "shot in the dark" type and are not taken on any periodic basis. They are best used for spot checks, as a quality assurance audit on other methods of sampling and analysis, or as an enforcement tool.

The type of discharge, whether it be air, surface water, a treatment discharge, or a groundwater sample, may also determine the type of sampling performed, whether continuous or individual.

Table 7.2. Field Sampling Considerations

Containers

- Glass:
  - Positive: organics may be preserved
  - Negative:
    - Breakage
    - Acid leaching of metals
- Plastics:
  - Positive: metals may be preserved
  - Negative: Organics may leach into sample
- *Never*
  - Rinse the containers with the sample!
  - Use PVC, ABS, styrene (adds organics to the sample)

Preservatives

- pH control:
  - Acids:
    - Control of microorganisms
    - Prevent precipitation — metals
    - Lower volatility — organics
  - Bases: CN and sulfide preservation
- Chemical addition:
  - Often specific for a given analyte:
    - Ascorbic acid (CN)
    - Zn acetate (sulfide)
  - Other complications possible with analytical method
- Refrigeration:
  - Nearly always used
  - Lowers volatility of organics
  - Retards chemical reactions

Other Considerations

- Holding times: critical for certain analyses
- Duplicate samples: usually the minimum
- Head space:
  - Gas phase partitioning possible
  - Without it, breakage possible if freezing encountered

Samplers

Some of the manual samplers that are used for sampling hazardous materials are seen in Figure 7.1. These include the "coliwasa" (composite liquid waste sampler), the weighted bottle sampler, the thief sampler, and the sampling trier. These are the recommended sampling tools for liquid and solid wastes under SW846. A variety of other methods including split spoon, bailer, auger or drilling samples, or soil gas sampling are commonly used to quantify the concentrations of hazardous materials in environmental samples.

ANALYTICAL METHOD

When one is dealing with the analytical method to be used, there is usually a right way, a wrong way, and an approved way. Therefore, one has to be careful to predetermine the purpose for which the sampling is to be performed before picking the actual chemical analysis method. For instance, if the sample is to be used in an enforcement proceeding, one would have to use government reference methods, whereas if one was concerned with designing a treatment process, one could easily use one of several other methods.

There are many reference methods, some of which are published by government agencies such as the EPA, OSHA, or various state agencies. Other reference methods (many times incorporated by reference into the regulations) are those published by nonregulatory bodies which have, through the years, come to represent a certain high level of standardization. These are methods published by the American Society for Testing and Materials (ASTM), the American Water Works Association (AWWA), or the American Public Health Agency (APHA).

Other reference methods may have no particular legal standing other than the fact that they are in common practice. Again, the choice of test methods depends on the purpose of the data.

For solid and hazardous waste the key document is EPA's SW846. It is the standard reference method for hazardous waste.

Water discharges under the Clean Water Act, however, are analyzed using the methods finalized in 40 CFR Part 136.[5] These include biological and inorganic test methods, with cross-references to EPA, ASTM, and the Standard Methods for the Examination of Water and Wastewater. Other methods under the Clean Water Act include those for nonpesticide organics, pesticides, and radiological materials.

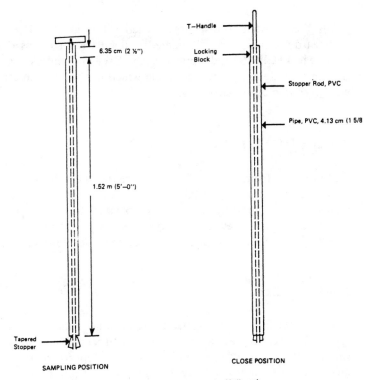

T—Handle

6.35 cm (2 ½")

Locking
Block

Stopper Rod, PVC

Pipe, PVC, 4.13 cm (1 5/8

1.52 m (5'—0'')

Tapered
Stopper

SAMPLING POSITION

CLOSE POSITION

Composite liquid waste sampler (Coliwasa).

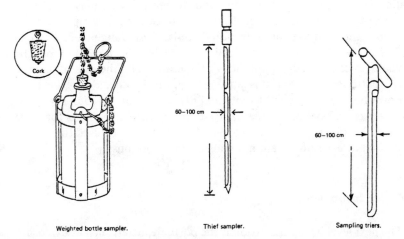

Cork

60—100 cm

60—100 cm

Weighted bottle sampler.   Thief sampler.   Sampling triers.

Figure 7.1.   Liquid (coliwasa and bottle) and solid sampling tools.

Test methods for air discharges, whether they are NESHAP emissions or hazardous compounds regulated under RCRA such as dioxins, are located in 40 CFR 60 and Appendix A to SW846. Readers are referred to the appropriate agency to determine which method to use for sampling any given contaminant.[6]

ANALYTICAL TECHNIQUES

Quite apart from the specific lab test methods are the analyzers used in environmental analysis. Analytical techniques range from field instruments to transportable instruments, to vans, to laboratory analytical instrumentation. They differ primarily by the amount of "sample workup" required, the cost involved, and the degree of accuracy.

Hand Instruments

Hand-held instruments generally require no sample preparation and are used to determine if a hazardous constituent exists in a gas or water sample. They give an order of magnitude or approximate contaminant level but are sufficient only for general indications or preliminary information.

The colored detector tube may be used for a gaseous sample concentration when one knows the compound of interest. One may take these "length of stain" indicating tubes to a site, draw the gas sample through the tube, and read the concentration directly. Costs range from $2 to $3 each for the tubes and from $200 to $400 for the hand pump. The advantages of this include simplicity of operation and the lack of moving parts, batteries, or hazardous waste samples to handle.

For water samples, conductivity and pH may be analyzed onsite using an analyzer approximately the size of a pocket flashlight. Costs for these instruments range from $100 to $1500. Spot tests may be used to indicate only the presence of certain minerals or common ions.

Portable Instruments

The next level in sophistication is the portable meter, such as explosion meters, oxygen meters, turbidity meters, or other gas or water phase analyzers. These measure some unique property of a contaminant, such as light absorption or heat capacity. They may range in cost from $500 to $3000.

Transportable Instruments

Transportable instruments or lab instruments mounted in a van represent a higher level of accuracy. These techniques require little workup and usually are applied in spill response situations. They are generally limited to vapors or easily extracted liquids and are analyzed most commonly by a gas chromatograph. The lab generally requires its own power source. It does offer greater accuracy and the potential to more thoroughly identify a contaminant and its concentration. Costs may range from about $2000 to $20,000 or more.

Lab Instruments

The highest levels of precision are found in analytical laboratory instruments. In this situation, the sample must be taken to the instrument. Sampling container preparation, transport, and lab workup of the sample prior to analysis must be performed before the instrument ever "sees" the sample. In a modern analytical laboratory, there are a range of techniques that may be used for determining the concentration of a material in a sample. These range from the original wet chemical methods to sophisticated instrumental methods. Since these represent the highest quality data, they are preferred where cost is of lesser importance.

LAB METHODS

Just as the field of chemistry is divided into organic and inorganic categories, so are instrumental methods divided between those that analyze for inorganic species and those that analyze for organic species.

Inorganic Methods

The inorganic instrumental methods range from specific ion electrodes (which are very specific) to atomic absorption spectroscopy (AAS) that will identify only certain elements in either a flame, graphite electric arc, or cold vapor state. Only metals may be analyzed in an AAS technique. Therefore, carbon, oxygen, fluorine, sulfur, hydrogen, and nitrogen cannot be analyzed in an atomic absorption analyzer.

AAS will give only the concentration of an element with no differentiation between the chemical states of that element. An example is chromium. It may exist in a toxic Cr VI state with the much less toxic Cr III state in the same sample. If the test is attempting to distinguish between the two, an atomic absorption method would not be advised, since it does not differentiate between the two states of chromium; it would only give the total elemental concentration.

Another method used for elemental analysis is the inductively coupled plasma (ICP) method. While this method is able to analyze a number of elements simultaneously, it is less accurate than the atomic absorption method due to matrix interferences. However, the ability to analyze a number of elements at the same time is its greatest advantage.

Asbestos[7] is the only common hazardous waste analyzed by a microscope. Optical counting of asbestos fibers collected on a filter is commonly used for screening analyses. The forms of asbestos may be identified using transmission electron microscopes (TEM).

Organic Methods

Organic chemical analyses are generally handled by instrumental methods. The instruments representing the three most popular methods are the gas chromatograph, the liquid chromatograph, and the gas chromatograph/mass spectrometer (GC/MS).

The gas chromatograph (GC) is an extremely versatile tool and has been available for a number of years; however, it is limited to volatile compounds only (i.e, those that will vaporize in the gas phase without breaking down at temperatures up to 300°F). It also suffers from being a separation technique; that is, it does not form a specific fingerprint or identification of an organic compound. If, however, one knows roughly which compounds are present in a sample, one may utilize the separation time period (retention time) to identify the chemicals present. The retention time is a function of the separation column used and gas parameters such as temperature and flow rate. If used in conjunction with a confirming GC column, identification and quantification of species is more easily assured.

This separation technique must use some type of detector for the compound of interest. The photoionization detector (PID) is one in which an ionizable compound is detected in the instrument after it has been separated in the gas chromatograph and is "seen" by its ionized fragments which are induced by light energy. Other sensors include the Hall detector, the flame ionization detector (FID), and the electron capture detector (ECD). The latter is most often used for pesticides or chlorinated compounds.

Liquid chromatography operates in much the same fashion as a GC separation technique. However, it occurs in the liquid phase. Consequently, it has only limited application in environmental work. It is used for those compounds that may decompose at the high temperatures encountered in a gas chromatograph. Applications of a liquid chromatograph would be for compounds such as polynuclear aromatic hydrocarbons (PAHs).

The GC/MS system is the most powerful (and the most expensive) of all of the instrumental organic methods currently being used. It is really two instruments in one. In this system, the front half (GC) separates the compounds and then the mass spectrometer identifies what type of organic compound it is. The mass spectrometer operates by breaking the compound down into various ions which are then counted by a magnetic amplifier to form fingerprint analyses of the products by atomic mass. This system operates at high vacuum and requires precision electronics. Usually such a system is coupled to a computerized data base system that scans the field, compares the fingerprint analysis, quantifies it, and prints or stores the data.

If one is working with a particularly "dirty" sample with a large number of interferences present, it may only be possible to run one or two samples per hour. Also, one may be limited by the number of samples that may be run in any given shift in the analytical laboratory due to the need to clean up the chromatographic columns.

REFERENCES

1. "Test Methods for Evaluating Solid Waste — Volume 1A: Laboratory Manual, Physical/Chemical Methods," 3rd ed., U.S. EPA, SW846, November 1986.
2. 40 CFR 260.11.
3. Gilbert, R. O. *Statistical Methods For Environmental Pollution Monitoring* (New York: Van Nostrand Reinhold Company, 1987).

4. Taylor, J. K. *Quality Assurance of Chemical Measurements* (Chelsea, MI: Lewis Publishers, Inc., 1987).
5. 40 CFR 136.3, and Appendix A to Part 136.
6. Harris, J. C., D. J. Larsen, C. E. Rechsteiner, and K. E. Thrun. "Sampling and Analysis Methods for Hazardous Waste Combustion," EPA-600/8-84-002 (1984).
7. "Guidance for Controlling Asbestos-Containing Materials and Buildings," U.S. EPA Office of Pesticides and Toxic Substances, EPA 560/5-85-024 (June 1985).

8   MANAGEMENT

INTRODUCTION

The sound management of hazardous materials is the first step in health risk reduction. Management practices, when carried out diligently, may attain a "near zero" discharge of hazardous materials. Also, public opinion and Congressional intent are pressuring the industrial sector to minimize (if not eliminate) use of hazardous materials and generation of hazardous wastes. Where they are not eliminated at the source, hazardous materials and wastes must be treated so as to diminish impacts on the environment. In this chapter, we will deal with management of hazardous materials as applied to a typical generator or user.

The primary burden of hazardous materials management falls upon the generators and users of hazardous materials. Therefore, it is important that companies dealing with hazardous materials have an ongoing hazardous materials management strategy.[1] A proactive strategy of management would have two goals: first, to handle hazardous materials safely, and second, to minimize costs due to improper management.

As a cost of doing business, hazardous materials management must be an ongoing part of any company's business plan. This includes all aspects of regulation: worker health and safety, generation, transportation, treatment, storage, and disposal. Hazardous materials management thus should receive the high priority that sales and profits receive. These must be in balance, since improper management can have adverse effects on both sales and profits.

In order to be cost-effective, any industrial hazardous materials management strategy must be a process-specific activity. Generally speaking, processes involving metal cleaning, painting, stripping, manufacturing, treatment of hazardous materials, storage, and spills are involved. And each has corresponding hazard reduction activities, such as recycling or reusing materials, substituting other materials, or minimizing waste production. These will all help meet the goals of handling hazardous materials safely and minimizing costs due to improper management.

Hazardous materials management today has two basic modes of operation: routine activities and those that may involve an emergency or a release of hazardous materials to the environment.[2]

The initial focus of this chapter will be on routine management activities. This will include a review of corporate policy and integrated management systems, and will provide a review of specific waste minimization approaches. Emergency actions for which any handler of hazardous materials must be prepared will be considered later in the chapter.

PROACTIVE HAZARDOUS MATERIALS MANAGEMENT

Because hazardous materials are regulated in transport, discharge, storage, treatment, and disposal, corporate management attention must be given to these activities. Also, the effects of potential bad press, accidents due to improper management, and the cost of doing business must all enter into the thinking of a top-level corporate management approach. The cost of failure to consider these or the costs of spill cleanup are definite incentives.

Corporate Policy

In order for the twin goals of handling hazardous materials safely and minimizing costs to be realized, successful management of hazardous materials must work from the top down.

Thus, when a facility environmental manager is required to maintain compliance with hazardous materials regulations, a written corporate policy statement is essential. With a corporate policy, the objectives of a proactive hazardous materials management strategy will be easier to accomplish. Figure 8.1 illustrates the different interests that are impacted by corporate policy.

The elements that could be expected for a successful policy on hazardous materials management will consist of a number of different requirements, including a commitment by the corporation to comply with or exceed all federal, state, and local laws, rules, and regulations.

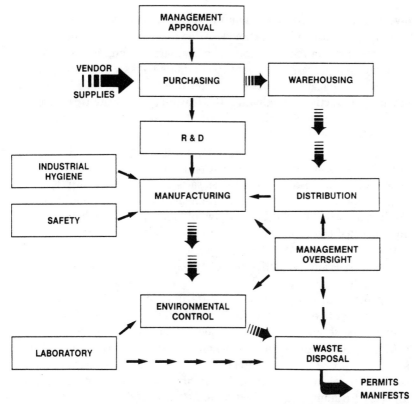

Figure 8.1.   Corporate interests impacted by hazardous materials policy.

A mandate for a hazardous material tracking system should be included. Such a tracking system would maintain a facility's inventory and would track hazardous material from the time it reaches the receiving dock until it leaves the facility—either through a permitted environmental release or as a manufactured product. Typical of such an activity would be a mass balance of hazardous materials throughout the facility, including bar coding of received materials, inventory of waste drums by number, and tracking of quantities of materials going into specific manufactured goods on a per-unit basis.

A commitment to active pursuit of waste minimization at each facility should be a part of a policy statement. Certainly, the largest return on an investment is realized by eliminating waste problems at the source. Specific types of waste minimization activities are noted later in this chapter.

A recommendation to establish a technical committee to evaluate materials and processes for ways to eliminate risks by substituting safer materials for hazardous ones or by reducing the generation of waste is a key in-house activity. Evaluating the feasibility of onsite treatment for residual waste materials that are hazardous is another active part of a policy statement.

Having the environmental staff reporting separately to the highest levels in corporate management is a policy that would elevate environmental concerns to the same level as sales or profits and keep the concerns of hazardous materials management from being subordinated to production activities. Such a policy would include the staffing of at least one qualified environmental control professional.

Requiring senior plant management officials to establish waste minimization goals and evaluating those officials' performance (and bonuses) is an excellent approach to accomplishing waste minimization. Incorporating the total cost of handling hazardous materials and hazardous waste as a part of each division's profit and loss statement is another activity that should be actively pursued by policy statement.

Mandating personnel environmental awareness training should also be an active part of a policy statement, as well as establishing a proactive compliance review program (on a monthly basis at a minimum). Certain aspects of training are required by RCRA and OSHA.

In addition, requiring procedural coverage for environmental matters in terms of waste handling, equipment operation, inspector visits, and product development should all be part of the approach. Incorporating these activities in a three-ring binder "cookbook" would help to maintain up-to-date proactive stances with respect to hazardous materials.

Establishing a centralized environmental control recordkeeping system for the plant and at the corporate level would provide the necessary feedback on current status of environmental compliance.

Environmental reviews should involve corporate transactions, environmental audits, and activities such as acquisitions, divestitures, leases, mergers, and lending operations.[3]

Obtaining competent legal counsel is a must, as is a corporate commitment to eliminating dependence on land disposal of hazardous waste.

Implementing an Environmental Control Policy

A proactive stance in hazardous materials management policy includes five things: coordination, dedication, time, planning, and money. Coordination of hazardous materials management is vital between divisional staff, corporate staff, production staff, the legal department, and the technical and engineering staff that deal with hazardous materials.

Figure 8.2 illustrates the different aspects of a hazardous materials/ waste management system that should be considered in implementing a proactive corporate policy.

Time and thorough planning are essential parts in a proactive hazardous materials management strategy. The plan should include prioritizing the actions to be taken, briefing of management and organizational staff, a phased implementation, provisions for program flexibility, and establishment of a baseline of activities. The baseline should inventory the hazardous materials currently in use (both in stock and projected); provide a definition of the processes that use those materials, preferably on a "per unit item of production" basis; and finally, document the existing management practices for each of those materials.

The actions to be taken to deal with hazardous materials must be prioritized with respect to level of risk, ease of remedy, and lowest cost to achieve the result needed. Management support, communication, and coordination are to be required at all levels of the organization. Everyone must understand that the twin goals of safety and minimizing cost are all part of the requirements of hazardous materials management.

Of course, any program that is well planned and implemented with respect to hazardous materials management will cause increased operating costs. These costs are short-term and should be considered insurance against future liability.

A helpful management tool is the waste minimization audit. This is not the same as an environmental audit, since it focuses on industrial processes. Because it deals with specific production at the source of waste generation, this effort can be very cost-effective. Other management tools include employee education, training, and monitoring activities; program and procedure review activities; and studies of the economics of the process, recycling activities, and reutilization of hazardous materials.

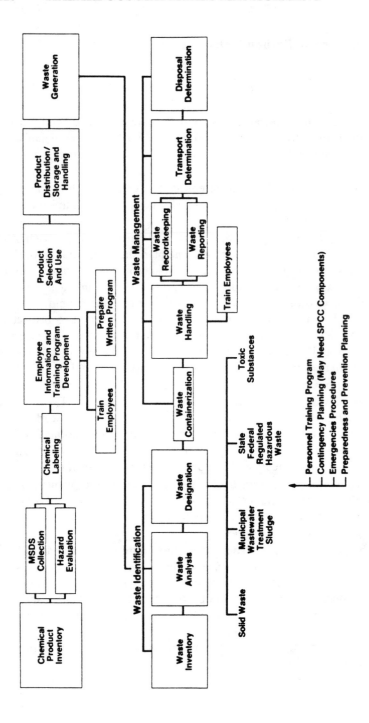

Figure 8.2.   Key elements in a corporate hazardous materials management system.

Alternatives to Hazardous Materials Generation

The most powerful management tool for reducing the risks of hazardous materials is reduction of the quantities of hazardous materials used. Reduction activities fall into two broad patterns. These correspond roughly to activities that can be undertaken onsite and those which can be instituted as offsite measures.

Offsite Waste Minimization

Offsite measures that help reduce waste volumes include waste exchanges and recycling of waste materials or waste products.[4]

Catalogs and newsletters that provide listings of recyclable hazardous wastes as well as technical assistance on waste reduction, economic incentives, information transfer, and regulatory incentives to reduce waste are often obtainable from state agencies. Listings may be organized by location or by type of material in the statewide inventory. The latter are usually categorized by waste type (solvents, spent products, acids, etc.) and by the quantity available. Transactional, financial, and transportation details are left to the parties involved.

Listings of commercial recyclers by type of material they will accept and the restrictions upon receipt of such materials may also be available. These typically involve acids, precious metals, oils, and solvents that have some intrinsic value other than as fuel. Of these, waste oils and solvents form the majority of materials that are recycled offsite.

Onsite Waste Minimization

Since the treatment and disposal of hazardous waste is complex, time-consuming, and costly, a reasonable first step would be to try to eliminate or minimize hazardous waste generation at the source.

This becomes a part of a generator's management program and will generally include a variety of waste minimization approaches. Among these will be attempts to:

- reformulate materials to use lesser amounts of (or eliminate) hazardous materials
- substitute less hazardous or less toxic raw materials
- recycle materials
- reuse by-products or wastes

Onsite waste minimization activities are wide in scope. These activities include materials substitution as well as the broader category of waste minimization (which includes reuse, recycling, and modifying processes).

Materials substitution has the potential to eliminate worker exposure to toxic elements and has the positive cost benefit of reducing production time lost due to worker illness. However, substitution of materials is product- and process-specific, particularly when using military specifications for coating or plating activities. Materials substitution is better for performance-based specifications than for design-based specifications.

Where an environmental assessment of a process is performed and it appears that a less toxic material can be substituted, there are decided cost savings, since processes or hardware are less likely to be changed. Process changes, however, may provide better overall economics and should not automatically be rejected.

The following sections provide some general principles that may be utilized in a waste minimization framework. Specific examples are given that appear to hold the most promise for waste minimization and risk reduction.

Coatings and Stripping

Coatings modifications fall into three major categories: minimizing the amount of coating used, minimizing or eliminating the thinner or cleanup solvents associated with coating operations, and substituting one type of solvent for another in the coating formulation.[5] Stripping modifications include reducing or eliminating the toxic chemicals used to chemically dissolve coatings.[6]

Coating Modifications

Powder coating techniques. Powder coating technology, also called dry powder painting, is one of the major advances in the application of coatings. This technique is based upon the deposition of specially formulated heat-fusible powders on metal substrates. Since no solvents are used in the system, pollution and safety problems associated with solvent-based coatings are eliminated. Cleanup solvents are greatly reduced. Air emissions of volatile organic compounds (VOC) are eliminated, as are thinners and leftover coatings.

In addition to the environmental advantages offered by dry powder coatings, the process provides technical production and cost benefits. Productivity is increased because without solvents the coating can be cured immediately after application. A technical advantage is that special coating materials may be applied by dry powder that could not normally be applied by conventional solvent-based systems. In addition, complex surfaces are more evenly coated in dry powder systems. For some applications, a single coating could replace multiple coating applications.

The electrostatic system is the most commonly used dry powder coating technique. In this application, dry powder is sprayed onto a surface that is electrostatically energized. The particles of dry powder are metered into compressed air through a special spray gun and sprayed on the surface. An electrode in the spray gun ionizes the air-and-powder suspension using direct current and the dry powder particles become charged. The substrate to be coated carries the opposite charge and the powder is electrostatically attracted to the surface and coats it. The particles are then fused to the surface and cured in conventional ovens. Any powder overspray is collected in a conventional air filtration system and reused. This also eliminates disposal of overspray solutions associated with liquid solvent-based paints.

The one major limitation in dry powder painting is that the items to be painted must be able to withstand curing temperatures of 350°F for 30 minutes. Aluminum alloys or very thin sections cannot be subjected to these conditions without significant loss of strength.

This waste elimination technique operates to minimize the amount of coating used and eliminates thinner, cleanup solvents, and wastewater.

Wet electrostatic techniques. Wet electrostatic coating is similar to the deposition of dry powder coatings by electrostatic attraction. It differs in that some solvents are used; however, the solvent content is much lower than with conventional spray painting. Overspray of organic compounds in the coating is minimized, resulting in significant hazardous waste reduction. Wet electrostatic coating is widely used for painting aircraft parts and other small, complex, nonaluminum metallic articles. There are, however, some concerns over potential safety hazards of imparting high voltage to a part that may contain fuel vapors. This technique minimizes waste by minimizing the amount of coating overspray.

Electrocoating. Electrocoating is somewhat similar to metal plating and is commonly used in automotive body coats. In this process, metallic or other electrically conductive parts are dipped into a solution that contains specially formulated ionized paint. The action of electric current induces the paint ions to deposit on the metal part. These paint formulations are a special class of waterborne nonvolatile organic compounds. Therefore, hazardous waste production is minimal and VOC air emissions are almost eliminated. The only limitation is that dip tanks are required; therefore, the size of items to be painted is limited. This technique minimizes hazardous waste by minimizing the amount of paint being used and eliminating overspray.

Water-based coatings. Water-based coatings are used extensively in industry and on a limited basis by the military. In water-based coatings (as the term suggests), the carrier solvent is a water solution rather than an organic solvent. Hazardous organic compound generation is less with water-based paints compared with solvent-based paints. The most significant decrease is in VOC emissions, which are nearly eliminated in water-based paints. In addition, the use of solvents for cleanup is eliminated, and no solvents are needed for paint thinning. Wastewaters generated from water-based painting contain fewer toxic organics (alcohols and ketones are used rather than aromatic and halogenated compounds) and lower percentages of solvents are in the formulation. In industry, water-based paints are normally used in applications requiring only moderate protection or where decorative requirements are of prime importance.

There are two key disadvantages of water-based paints. First, the surface must be completely free of oily films or the paint will not adhere. Second, water-based coatings require longer drying times or even oven drying in cold or humid weather; this requirement may result in significant expense to retrofit an installation for water-based paints. This approach minimizes waste by substituting water for an organic solvent.

High-solids coating. High-solids coatings, which are similar in composition to solvent-based coatings, are becoming more widely used for some industrial applications. High-solids coatings contain about 25% to 50% solids and, compared to solvent-based coatings, use lower molecular weight paint resins with numerically more reactive sites to aid in coating polymerization. The finished coat is comparable to normal solvent-based coatings.

The high-solids coatings require special equipment using heated spray guns and nozzles for application because of their higher viscosities. Because less solvent is used, less is available to "wet" metallic surfaces contaminated with oils; therefore, surface preparation is more critical. Spray application is also more difficult because there is a tendency to apply too much coating to achieve a similar "wet" appearance obtained in normal solvent coatings. In the industry, the general opinion is that high-solids coatings will become the "standard" to replace normal solvent-based coatings. The major advantage of this technique is the ability to comply with the VOC limitations while using nearly the same equipment and application techniques.

Improved painting techniques. Other changes that minimize waste include those that improve the transfer efficiency of the paint to the part to be coated. Among these are combinations of air-assisted and airless spray applications, and conveyorized paint systems that allow parts to be painted, rotated, and painted again without being touched. The potential for utilization of robotics with certain conveyorized systems is proven technology. In general, each of these operates by minimizing the amount of paint used and in some cases minimizes the amount of thinner and cleanup solvents through more effective transfer efficiencies.

Paint Stripping

Paint stripping is the process of removing paint and coatings from surfaces in preparation for recoating or resurfacing. Complete stripping is often necessary before a new paint or coating will properly adhere to the existing surface.

Alternatives to current solvent and chemical paint stripping techniques are available that reduce the generation of hazardous waste but require new equipment. These alternatives include plastic media blasting, laser paint stripping, flash lamp stripping, water jet stripping, and dry ice pellet stripping. Of these, the plastic media blasting appears to have the most promise for current applications.

Plastic media blasting is a replacement for sandblasting in which plastic particles or flakes are shot against the surface of the old painted part. Since the material used is softer than sand and tends to rebound (and not be broken up), it is somewhat more attractive than sandblasting itself.

Through careful control of the size and physical properties of the pieces and the conditions of the process, the plastic media can be separated from the loosened paint particles and recycled. Generation of wet hazardous waste (solvents, paint sludge, and water) is completely eliminated. A small volume of dry waste is produced that may be classified as hazardous because of the metals content of the original paint.

The two key parameters for successful use of plastic media blasting are its hardness and reusability. First, the paint must be softer than the plastic media, which, in turn, must be softer than the surface underneath the paint coat. Second, the plastic pieces must be durable enough to be reused over and over to minimize the amount of waste residue. This technique has been shown to be effective in stripping and removing a wide variety of coatings from a number of substrates. However, care must be exercised on composite surfaces, thin-skinned aluminum, and other fragile materials. In particular, problems with unraveling of composite fiber substrates have been encountered in blasting composite surfaces that do not have a resin-rich surface. Also, small or complex surfaces may not be "reachable" by plastic media blasting.

Solvent Recovery and Reuse

The most promising technology identified that can be applied to almost all industrial facilities is organic solvent distillation and condensation. Alternative technologies other than distillation that may also be used, depending upon a particular process, are centrifugation, filtration, ultrafiltration, reverse osmosis, and activated carbon.

Distillation. This technology relies on heating a solvent sufficiently to vaporize it and condensing the vapor. The condensed vapor is then reused as the recovered material. If the boiling point of the solvent is high enough, the distillation can be accomplished under a vacuum to lessen thermal decomposition of the solvent.

The waste materials, called still bottoms, must be disposed of as a hazardous waste. However, volume reduction factors of up to 15:1 can be found. There are many commercially available self-contained distillation systems that can distill solvent quantities at rates ranging from one-half to 100 gallons per hour. Off-the-shelf units can be purchased and installed in any sheltered area that has electrical power and cooling water available. These types of systems can be used most effectively when there is a single-component system being used. With multiple-solvent systems, distillation is not advised.

Physical methods. Technologies such as centrifugation, filtration, ultrafiltration, and reverse osmosis are primarily used to separate phases in a waste stream. These technologies are appropriate for emulsion-type organic fluids used in machining processes, but are not usually recommended for recycling cleaning solvents.

Centrifugation and filtration are used to remove metal particles generated by machining processes to allow the machining fluid to be reused. Ultrafiltration and reverse osmosis are used primarily to separate water from emulsified water-oil streams. These processes are specific to certain waste streams and are not commonly used.

Activated carbon. Activated carbon is usually used to capture airborne solvents. The airborne solvents are adsorbed onto the surface of the carbon, and then steam is used to strip the adsorbed organics off the carbon. The solvent can then be recovered from the steam condensate. This approach is limited to single-phase systems and solvents immiscible with water (such as perchloroethylene).

Metal Plating Waste Minimization

Technologies to reduce the amount of hazardous waste from metal plating operations are available and show a great deal of promise. Process changes may require some capital outlays due to equipment modifications. Modifications include improved housekeeping, reducing "drag-out," modifying rinsing, recovering metals from rinsewaters, reducing or eliminating tank dumping, and changing to less hazardous plating materials.[6,7]

Housekeeping

This approach requires little or no capital investment and can realize significant savings in raw material usage and wastewater treatment. Among the housekeeping items would be:

1. repairing leaking valves, tanks, and pumps
2. inspecting tanks, etc., to avoid failures that result in dumping of the entire solution
3. installing high-level alarms for overflow
4. maintaining racks and anodes to prevent contamination of baths
5. minimizing the volume of water used during cleanup operations
6. training personnel to understand the importance of minimizing both bath contamination and wastewater discharge
7. properly cleaning and rinsing parts prior to plating to minimize contamination of the plating bath

Drag-Out Reduction

The shape and design of the parts and racks can significantly affect drag-out rates as well as operations.

Modifications should first be implemented to reduce bath dumps before concentrating on reducing drag-out. A more favorable rate of return is realized by implementing drag-out reduction techniques at decorative chrome, cadmium, and zinc plating lines, where plating times are relatively short and drag-out is significantly greater than in hard chrome plating.

Drag-out can be reduced by decreasing either bath viscosity or surface tension. Viscosity can be decreased by reducing the chemical concentration of the bath or by increasing temperature. Surface tension can be reduced by either adding nonionic wetting agents or increasing bath temperature. These modifications improve the drainage of plating solutions back into plating baths or reduce the concentration of metal in the drag-out. Lowering the velocity of withdrawal of parts from a bath can dramatically reduce the thickness of a drag-out layer because of surface tension effects.

Drag-out can be captured by the use of drain boards, drip bars, and drip tanks and can be returned to the bath. These simple devices save chemicals, reduce rinse requirements, and prevent unnecessary floor wetting. Significant drag-out reduction can be accomplished if platers carefully rack and remove parts so as to minimize entrapment of bath materials on surfaces and in cavities. Parts should be designed to promote drainage.

Air knives can be used to knock plating films off parts and back into process tanks. This technique is particularly effective in removing ambient temperature solutions from plated parts. Spray rinses are also effective in removing drag-out from parts. The part is held over the plating tank and sprayed with rinsewater. Over 75% of plating chemicals drain back to the plating bath. Spray rinsing is best suited for flat parts that are hard chrome plated since evaporation rates in these baths can exceed rinsewater requirements.

Rinsewater modifications and reductions in rinsewater flows may not reduce the amount of toxic metals to dispose of, but can they reduce the volume of liquid waste that must be processed in wastewater treatment systems. If rinse flow rates are reduced sufficiently, it is possible to use rinsewater for evaporative losses (make-up) in the plating tanks. Reducing flows can also increase the efficiencies that are recovered for processes such as ion exchange, reverse osmosis, and electrodialysis.

Chemical Recovery Processes

Evaporation, reverse osmosis, ion exchange, and electrodialysis have been used to recover chemicals from rinsewaters in the metal plating industry. These processes reconcentrate plating solutions from rinsewater and produce a relatively pure water that can then be reused for rinsing. Both general and site-specific factors must be evaluated to determine the recovery process that is best suited for a particular plating operation.

Evaporation. Evaporation may be used to recover plating chemicals from rinse streams. In this process, enough rinsewater is boiled off to concentrate the solution sufficiently to return it to the plating bath. The steam can then be condensed and reused for rinsing. Because of their high energy use, evaporators are most cost-effective in concentrating rinsewaters that are returned to hot baths such as those used in chrome plating, where high evaporation rates increase the concentration.

Reverse osmosis. Reverse osmosis (RO) is a demineralization process in which water is separated from dissolved metal salts by forcing the water through a semipermeable membrane at high pressures. The key components of an RO unit are the membrane, the membrane support structure, a vessel, and a high-pressure pump. Rinsewaters must be filtered to prevent fouling of the membrane. RO units concentrate most divalent metals (Ni, Cu, Cd, Zn) from rinsewaters to a 10% to 20% solution by weight. The key item in effectiveness of an RO unit is the membrane; therefore, process and operating uncertainties can significantly affect cost of such an approach.

Ion exchange. Ion exchange uses charged sites on a solid resin matrix to selectively remove either positively charged ions or negatively charged ions from a solution. Ions removed from the solution are replaced by an equivalent charge of other ions from the resin, thus the name ion exchange. Exchanged rinsewater is normally recycled.

In metal plating operations, anionic exchange resins have been used to recover chromic acid from rinsewaters, typically exchanging hydroxide ions for the negatively charged chromic acid anions. An ion exchange system typically consists of a wastewater storage tank, prefilters, exchanger vessels, and caustic or acid regeneration equipment.

In general, ion exchange systems are suitable for chemical recovery applications where the rinsewater has a relatively dilute concentration of plating chemicals and where a relatively low degree of concentration is required for recycling of the concentrate.

Ion exchange has been most successful in recovering chromic acid and nickel from rinsewaters, but some problems have been encountered in concentrating mixed metal solutions. By using the ion-exchanged water for rinsing, fresh water consumption can be reduced by 90%. However, ion exchange recovery systems are not cost-effective when drag-out rates are low. In addition, waste regenerative brine from the resin can be difficult to dispose of and expensive to treat.

Electrodialysis. Electrodialysis concentrates or separates ionic species in a water solution through the use of an electric field and semipermeable ion-selective membranes. Applying an electric potential across a solution causes a migration of cations toward the negative electrode and migration of anions toward the positive electrode. These units are packed with alternating cation and anion membranes between alternating cells. Typically, these units may contain anywhere from 10 to 100 cells for enhanced efficiency.

Electrodialysis has been used to recover cationic metals from plating rinsewaters. Unlike ion exchange and reverse osmosis, the maximum concentration limit of an electrodialysis unit is limited only by the solubility of a compound in solution. A major disadvantage of electrodialysis is that all ionic species are nonselectively removed. Therefore, plating baths may build up levels of impurities which would cause the bath solutions to become contaminated. Specific applications and processes must be analyzed prior to utilization of electrodialysis.

PROACTIVE EMERGENCY PLANNING

The second major type of risk reduction activity that a company using hazardous materials must face is emergency planning.[8] The key item in management of hazardous materials when it comes to emergency response is *planning in advance*.[9] Also, current laws require that planning be performed well in advance of any type of accidental release.[10] Title III of SARA[11] has extended a number of emergency notification and planning requirements to all states and territories. Local and state laws may be even more rigorous.

Under these laws, every state and all businesses that handle greater than minimum quantities of hazardous materials must develop comprehensive emergency response plans. In addition, all businesses must report a significant release of hazardous materials. In conjunction with state agencies, regulations have been developed for minimum standards for local area plans and minimum standards for business plans. All plans must include provisions for immediate notifications, mitigation of an actual or threatened release of a hazardous material, and evacuation of an area. Area plans are formed by governmental planning committees.

Accidental releases that result in emergency response can be of two types: those that involve onsite spills or accidents, and those that involve offsite spills or transportation accidents. Offsite accidents will not be considered in this section. Onsite releases or spills form the basis for most emergency plans.

Key Provisions for Emergency Plans

Provisions mandated for business and governmental emergency plans are outlined below. State laws generally follow these approaches.

Business Plans

- Business plans must preidentify hospitals and other medical resources that have capabilities to provide treatment for hazardous materials accidents and exposures.
- Businesses must include procedures in their emergency plans to reduce the consequences of a release of a hazardous material; that is, a handler must identify actions that will mitigate, prevent, or abate hazards.
- Procedures must exist to inform employees at a facility and the surrounding public of a release of a hazardous material.
- Business plans must include programs for initial and refresher training of appropriate employees in:
  1. the safe handling of hazardous materials used by the business
  2. methods of working with the local public emergency response agencies
  3. the use of emergency response resources under control of the handler
  4. other procedures and resources that will increase public safety and mitigate a release of hazardous materials

State and Local Plans

- Area plans must include provisions for the notification of, and coordination with, the appropriate response entities including law enforcement, fire suppression, public health, medical response, hospitals, incident control and mitigation, and reception and care of evacuated persons.

- An area plan must contain provisions to identify and utilize secondary communications systems such as amateur radio services and cellular car telephones.
- Preincident planning must include a clear delineation of the responsibilities of the significant response entities for various functions such as site security, fire suppression, and evacuation.
- Area plans are required to contain provisions for training emergency response employees in 10 specific areas:

    1. Emergency procedures — response personnel must receive training in hazard identification, safe approach to the scene, proper handling of hazardous materials, and a variety of potential complex incident scenarios.

    2. Health and safety procedures — training must be provided in safe approach, recognition, and evaluation of hazardous materials, and proper monitoring and decontamination procedures with training tailored to the hazards present in a particular site.

    3. Equipment use and maintenance — emergency response employees must be trained in the proper use of specialized equipment required to respond to hazardous materials.

    4. Mutual aid — response organizations must be trained to effectively identify and obtain resources from other jurisdictions.

    5. Medical resources — medical facilities that can properly maintain and administer specific treatment for hazardous materials exposure and contamination must be identified, as well as resources outside the area that can provide the essential information or treatment.

    6. Evacuation — response personnel must be familiar with specific evacuation protocols such as the distance for security perimeters, evacuation routes, possible release pathways, and reception and care of evacuated persons.

    7. Monitoring and decontamination — emergency response personnel must be trained to recognize, safely remove, and contain hazardous materials.

    8. First aid — emergency response personnel will be required to undergo training in basic first aid for injuries from hazardous material accidents in order to stabilize victims until a higher level of medical treatment is available.

    9. Public information — training is to be provided in procedures

for making necessary safety information available to the public.

10. Psychological stress — response organizations will be required to provide training to reduce psychological impacts of a release of hazardous materials.

Problem Areas and Information Sources

Some of the major problem areas that receive the bulk of the attention in these types of plans are communications, public relations, preplanning in the event of breakdown of infrastructure or support systems, identification of the actual materials involved, and the boundary of the potentially affected area.

Among one of the major information sources that are to be utilized are the Material Safety Data Sheets (MSDSs). These are required by OSHA to be onsite, along with the plant-specific inventory of hazardous materials. The MSDSs for every hazardous substance give the identity of the hazardous ingredients, the physical and chemical characteristics of the substance, the physical hazards, the reactivity and health hazards involved, and safe handling and use precautions. In addition, the control measures to reduce harmful exposures are also listed in every MSDS.

Appropriate preparation of contingency plans must be followed by "dry runs" to check the adequacy of each element of the plan. Difficulties encountered must be resolved before the "real thing" occurs. Active reviews and audits of these plans and the dry runs will help to keep plans current and will keep management, staff, and plant personnel up to date. They will also point out any needs for additional training, communication system changes (such as using cellular car telephones rather than "hardwire" electronic systems), or emergency equipment. An annual review of all procedures is generally recommended.

Having accomplished hazardous materials reductions, implemented waste minimization strategies, and prepared for emergencies, a proactive hazardous materials manager will have come full circle to minimizing risks to human health. These practices, if maintained, will also accomplish the objectives of minimizing costs and handling hazardous materials safely.

REFERENCES

1. Frances, M., Rockwell International. Personal communication (December 3, 1986).
2. *Hazardous Waste Management Guide for the Small Quantity Generator* (Westminster, CA: California Safety Council, 1986).
3. Griffin, R. D., and P. Kaplow. "Hazardous Waste in Real Estate: Engineering and Legal Concerns," UCLA Extension, Department of Engineering and Science, Course #815.10 (May 2, 1987).
4. "Alternative Technology for Recycling and Treatment of Hazardous Wastes — Third Biennial Report," California Department of Health Services, Toxic Substances Control Division (July 1986).
5. "Solvent Waste Production Alternative Symposia Conference Proceedings," California Department of Health Services (October 1986).
6. Higgins, T. E. "Industrial Processes to Reduce Generation of Hazardous Waste at DOD Facilities (Phase 1 and 2 Report)," U.S. Department of Defense Environmental Leadership Project and U.S. Army Corps of Engineers (July 1985).
7. Peters, R. W. "Heavy Metals Separations," *Environmental Progress* 6(2) (May 1987).
8. "Planning Guide and Checklist for Hazardous Materials Contingency Plans," Federal Emergency Management Agency, FEMA-10 (July 1981).
9. Bellomo, A. J. *Hazardous Materials Incident Response* (Denver, CO: National Environmental Health Association, 1982).
10. "Chemical Emergency Preparedness Program — Interim Guidance," Revision 1-9223.01A U.S. Environmental Protection Agency, Office of Solid Waste and Emergency Response (November 1985).
11. Public Law 99-499, Title III of the Superfund Amendments and Reauthorization Act (1986).

9   WASTE TREATMENT AND DISPOSAL

The major objective in waste treatment and disposal is to minimize the exposure of people to toxic or hazardous materials. However, before one attempts to treat or dispose of hazardous materials, the characteristics of the various streams must be known.[1] Among these characteristics are:

1. the generation rate of each stream, such as gallons/day, pounds/hour, etc.
2. the chemical and physical characteristics of each stream
3. the toxicological features of the waste or material stream

To assure oneself that this type of characterization is complete, engaging the services of an analytical chemistry lab is required. For instance, chemical characteristics must include knowledge of whether a stream is acid or alkaline, whether it contains dissolved metals or solvents, or whether the material spontaneously reacts with air or water.

Knowledge of the physical characteristics of a waste or material stream should include information on vapor pressure or flash point (if it contains solvents or organics), density, percentage of major components (water or oil), and the number of phases (solid, liquid, emulsion) present. This information will allow technical and management decisions to be made in a context of facts rather than assumptions.[2,3]

BASIC TREATMENT PRINCIPLES

Whether the chosen course is onsite or offsite management or a combination of the two, the technical, regulatory, and logistical questions that must be addressed are basically the same. This is not to say that the firm that sends all its waste offsite will face the same problems as the one that manages an onsite TSDF, but the same issues must be addressed by both.

Assuming that one has hazardous waste to manage, and knows who is to do the work, the next step is to decide what is to be done. This might be constrained by a number of factors, including technical ones. Non-technically trained members of the management team may not be aware of such limitations. (For example, they might not understand that thermal treatment of wastes does not destroy heavy metals.) Clearly delineating the general types of limitations during the decisionmaking process will not only assist in organizing thoughts, but will also assure the staff that no reasonable alternative has been overlooked.

Ultimately, however, there are only three basic principles that can be implemented for "treating" hazardous wastes. These are:

- detoxification (including destruction)
- volume reduction
- isolation of the material from the environment

Each of these will contribute to lower exposure, lower risk, and, therefore, fewer health effects. It is important to realize that no one technique is adequate for all waste streams.[4,5] One common technique, air stripping of water contaminated with volatile organics (which exchanges air pollution for water pollution), negates all three basic approaches or principles!

Even in implementing these principles, there remains the final management of residuals, since treatment cannot achieve a zero level of contamination. Such management includes the controlled emission of stack gases to the air, the discharge of treated effluents into surface waters, or solids disposal to land.

TECHNOLOGY CONSIDERATIONS

The five major types of waste treatment operations are:

1. physical treatment—processes that, through concentration and/or phase change, alter the hazardous constituents to a more convenient form for further processing or disposal.

2. chemical treatment—processes in which hazardous constituents are altered by chemical reactions. In most cases, this means haz-

ard destruction. In some exceptional cases, the resultant product or products may still be hazardous, although in a more convenient form for further processing or disposal.

3. biological treatment — technically a chemical treatment, but classified separately because of its already widespread application for treating wastewaters, both hazardous and nonhazardous.

4. thermal treatment — processes that use high temperature as the principal mechanism for waste destruction. (Obviously, this involves chemical reaction, too.)

5. fixation/encapsulation — processes in which the hazards of wastes are reduced by immobilization.

Physical Treatment

The physical treatment technologies for hazardous waste management vary widely. They may be used by themselves, or in conjunction with other processes (such as those involving chemical reaction). Usually, physical treatment is used to reduce the volume of material for further processing or disposal.

Liquid/solid separators include such devices as gravity-settling ponds, clarifiers, froth or gas flotation units, and filters. In some cases these devices are themselves sources of hazardous waste; for example, RCRA lists sludge from an oil refinery's gas flotation unit as a hazardous waste.[6] Dilute sludges from such a device are usually processed further by mechanical dewatering equipment, such as vacuum filters, centrifuges, belt presses, and filter presses. With mechanical dewatering, wet sludges are being viewed with increasing disfavor for land disposal since the processed sludges still contain moisture. Thus, it is likely that sludge-drying processes will become more popular as landfill requirements are tightened. Frequently these processes are used to facilitate incineration and/or disposal options.

Crushing and grinding operations have several specific applications to hazardous waste treatment. Many processes that generate such miscellaneous hazardous wastes as work gloves, empty drums coated with residual chemical films, or other containers, might employ crushing or grinding for volume reduction. The other principal use for this treatment technology is in those cases in which a solid material's particle size must be reduced for proper physical or chemical reactivity in subsequent treatment steps.

Most mass transfer processes have some application to hazardous waste treatment. Distillation and evaporation can be used to split a feed-stream into two or more components, with one containing most of the contamination. Stripping is applicable to streams contaminated with trace constituents. In particular, the air stripping of trace concentrations of volatile solvents from contaminated groundwater has been used successfully in a number of installations. Absorption systems are not commonly used for hazardous waste treatment, but frequently an absorption system is used as an adjunct to other hazardous waste processing systems, such as controlling residual vapor emissions.

Solvent extraction systems such as liquid-liquid, solid-liquid, or supercritical fluid have been widely promoted for hazardous waste processing. The removal of polychlorinated biphenyls (PCBs) or dioxin from soils has been investigated, but process difficulties and costs have restricted the application of solvent extraction systems, and their commercial use is rather limited.

Adsorption systems, both carbon and resin, are being applied commercially to waste streams containing low concentrations of hazardous constituents. Activated carbon systems have become so common that many vendors have units available off the shelf.

Chemical Treatment

The goal of chemical treatment is to destroy the hazardous constituents in the waste stream or to convert them into a more easily treated form. Since chemical reactions involve specific reactants under specific conditions, these processes are often used when only one substance or its chemical "family" is involved. When chemical treatment is applied to a mixed composition waste, interferences such as side reactions, impurity interference, catalyst destruction, or unexpected end products may result.

Neutralization is the most widespread chemical treatment process. It has been practiced for a number of years and involves mixing two or more products (acid and alkali) to obtain a neutral, nonhazardous salt plus water. Waste acids or bases may be used provided that the impurities do not create other hazards.

Precipitation is frequently used in conjunction with neutralization. For example, soluble copper in an acid solution can be effectively precipitated by adjusting the pH to slightly above neutral. Sulfides may be employed to remove heavy metals, while lime or calcium chloride is used to remove fluorides. Electrochemical processes are also used for removing metals from process streams before discharge.

Chemical oxidation or reduction processes can treat hazardous wastes. However, they are usually applied only to dilute solutions because of the high cost of the treatment chemicals. They have the additional disadvantage that the reactions tend to be incomplete unless run under precisely controlled conditions. Another problem is that the oxidized or reduced end products may have hazardous properties of their own. Reaction of PCB fluids with metallic sodium eliminates the PCB compound but leaves behind salt and biphenyl (an aromatic compound of two linked benzene rings).

Biological Treatment

Biological treatment is limited to wastewaters containing organic compounds. As long as the hazardous organic constituents are not present at a toxic concentration, excellent organic destruction can be achieved. Cyanide and phenol solutions (below the toxic threshold) are virtually destroyed in an activated sludge system.

Organic constituents that are biodegradable in water matrices are easily treated, but bioadsorption — the adsorption of materials onto a cellular biomass — also plays a role. Concentrations of oil, chlorinated hydrocarbons, and heavy metals may be dramatically reduced from influent to effluent of a water treatment system by bioadsorption within the activated sludge. Thus, further processing and disposal of the treatment sludge requires careful planning because of potential releases of adsorbed constituents.

Waste treatment in soil is a form of biological treatment using existing native bacteria. Landfarming of oil sludges from petroleum refining has been used for years, and is relatively effective.[7] In situ bioaugmentation — the injection of bacteria into soils for contaminant treatment — is promoted for the cleanup of hazardous waste sites containing organic wastes, but the effectiveness of this process is unproven, except on lab- or pilot-scale applications.

Thermal Treatment

Incineration has become the favored technical choice for disposal of hazardous wastes containing organic compounds because most organic materials are efficiently destroyed. Incinerators, as a class, have been used for decades for the destruction of various hazardous materials (principally gases and liquids).[8]

Any controlled combustion device, from an existing steam boiler to a high-technology plasma-arc, is suitable for hazardous waste destruction, provided that it meets regulatory requirements for waste destruction efficiency. Many incinerators are dedicated to hazardous waste destruction, and sizes range from a simple liquid-injection unit to an oceangoing incinerator ship.

The rotary kiln, with its high temperatures and long residence times, has been adopted as a state-of-the-art combustion device for "generic" RCRA hazardous wastes such as solids, drummed waste, sludges, liquids, slurries, pastes, and loose bagged material.

Again, incinerators are only good for organic waste destruction. Concerns for meeting legal requirements such as minimum destruction and removal efficiency (DRE), criteria air pollutant limits, control of acid gases, and the degree of hazard presented by the ashes containing toxic metals may also influence the treatment method.[9,10]

For organic wastes in a water phase or solution, a wet-air oxidation may be used. At normal operating pressures and temperatures, organic destruction is typically 85% to 95%. This limitation, plus high costs, restricts the general application of wet-air oxidation processes. A disadvantage is that the technology is limited to water solutions with less than 5% total organics.

Fixation and Encapsulation

Fixation and encapsulation processes for containing liquids, semiliquid wastes, or leachable solid wastes are currently being investigated for containment of residual materials that were once placed directly in a landfill. Most of this technology is relatively new, except for a few selected practices adapted from the nuclear waste disposal industry.

The simplest fixation process is sorption, whereby free liquid containing dissolved hazardous metals is sorbed until the final product becomes dry. Fly ash, lime, clay, vermiculite, and zeolite are typical sorption media.

More sophisticated sorbents that also involve pozzolanic (aqueous hardening) reactions are lime–fly ash and Portland cement. Pozzolan/Portland cement systems have been widely used for nuclear wastes and have also been applied to industrial wastes. Modifications of this technology are often necessary because some materials inhibit the binding action. Soluble silicates, selected clays, emulsifiers, or surfactants are used in certain applications to solve this problem.

Selection of sorbent materials involves tradeoffs among chemical effects, costs, and amounts required to produce a solid product suitable for land disposal. Selection criteria include pH of the waste, quantity of sorbent needed to eliminate all free liquid, compatibility or reactivity of the waste and sorbent, level and character of contamination that might be introduced in the sorbent, and chemical binding properties of the sorbent for specific contaminants.

Some materials immobilize the hazardous wastes by encapsulation rather than sorption or chemical reaction. Organic polymers, vitreous agents (such as glass), and asphalt are examples of such agents that are currently being promoted.

There are also proprietary inorganic, inorganic/organic, and organic agents that are marketed for both fixation and encapsulation. In these cases, specialized mechanical equipment is required for the encapsulation process to ensure complete mixing prior to hardening of the final product. These costs must be considered as well as the cost of reagents.

In the selection of both the stabilizing material and the mixing equipment, compatibility of the waste and the matrix (the sorption or encapsulating material) is a major consideration.

The most serious concern about fixation and encapsulation processes is the uncertainty about their long-term effectiveness. Based on both experience and inference about chemical character, inorganic agents appear to be more reliable than organic ones. The principal concern is biodegradation and leaching of hazardous constituents into groundwater.[11]

Table 9.1 gives a summary of various treatment options for a variety of wastes. The relative costs of some of these methods compared to simple landfilling are given in Table 9.2.

ULTIMATE RESIDUALS MANAGEMENT

Apart from dispersal into the environment, the ultimate disposition of hazardous wastes will be in a solid matrix in a landfill or dedicated residuals repository.

The function of a landfill or residuals repository is to permanently

Table 9.1 Waste Treatment Options

Treatment Method/Process	Aqueous Solutions Inorganic	Aqueous Solutions Organic	Organic Liquids	Sludges & Slurries	Solids
Neutralization	X	X		X	
Precipitation	X				
Solidification	X				
Stabilization	X			X	X
Distillation			X		
Extraction		X	X	X	
Steam stripping		X	X		
Adsorption		X	X		
Oxidation	X	X	X	X	X
Incineration		X	X	X	X
Emulsion breaking		X	X		
Drying			X		
Dewatering	X		X	X	
Landfill/landspreading				X	X
Landfarming		X	X	X	
Surface impoundments	X	X		X	
Wastewater treatment	X	X		X	

Table 9.2 Relative Cost of Waste Management Methods

Method	Relative Cost
Landfill	1.0
Landspreading	1.3
Landfarming	1.5
Surface impoundment	0.7
Stabilization	1.2–3.0
Dewatering	1.5–3.0
Oil-water separation	1.5–3.0
Neutralization	1.5–6.0
Oxidation	2.0–20.0
Steam stripping	4.0–10.0
Adsorption	3.5–10.0
Liquid incineration	4.0–8.0
Rotary kiln incineration	5.0–10.0

keep the hazardous materials from migrating back into the environment (i.e., surface water, groundwater, or air). The two basic approaches to ultimate land disposal are:

1. the "natural" system
2. the engineered system

The effectiveness of a natural system depends on local geology and meteorology to keep hazardous materials from migrating away from the repository. Geologic features may enhance effectiveness by providing a subsurface water barrier (normally clay) beneath the deposited material. Depending on the formation, the permeability of the "natural" layer may be as low as 10^{-7} cm/sec (1 inch/year). This feature inhibits the movement of water-phase hazardous waste constituents in the subsurface environment for long periods of time.

Meteorology plays a role in determining how much water may enter the landfill prior to a final cap being placed on the site. If the rainfall were low enough (< 4 inches/year) and evaporation high (> 100 inches/ year) it is doubtful whether moisture could seep down far enough to transport contaminants to the natural barrier.

The engineered system, which may be either above grade (storage vault) or below grade (landfill), depends on the design of the containing structure, including leachate collection systems, to keep hazardous materials from migrating into the groundwater. Design considerations must include provisions for surface cover, side walls, and bottom to withstand earthquakes, corrosion, floods, sabotage, incompatible mixing, heavy rainfall, and structural damage during waste deposition. Leachate collection systems consisting of synthetic impermeable layers with water/ leachate collection pipes, nets, and sump pumps (along with monitoring wells surrounding the property) in various combinations are necessary for any design.

In addition, RCRA requires that certain minimum construction standards be attained, such as double-synthetic liners with double-leachate collection systems over an impermeable clay soil layer.[12] The state of California goes further and requires that the clay layer beneath a repository be naturally occurring with a permeability of less than 10^{-7} cm/sec, thus requiring natural systems to augment the engineered design.[13]

REFERENCES

1. "Treatment Technologies for Hazardous Wastes," *JAPCA* (January–August 1986).

2. King, Y. H., and A. A. Metry. *Hazardous Waste Processing Technology* (Stoneham, MA: Butterworth Publishers, Inc., 1982).
3. Mackie, J. A., and K. Nielsen. "Hazardous Waste Management: The Alternatives," *Chem Eng.* (August 6, 1984).
4. Cope, C. B., W. H. Fuller, and S. L. Willetts. *The Scientific Management of Hazardous Wastes* (New York: Cambridge University Press, 1983).
5. Stoddard, S. K., G. A. Davis, H. M. Freeman, and P. M. Deibler. "Alternatives to the Land Disposal of Hazardous Wastes, an Assessment for California," Toxic Waste Assessment Group/Governor's Office of Appropriate Technology, State of California (1981).
6. 40 CFR 261.32 (K048).
7. Brown, K. W., G. B. Evans, and B. D. Frentrup, Eds. *Hazardous Waste Land Treatment* (Ann Arbor, MI: Ann Arbor Science Publishers, Inc., 1983).
8. Bonner, T., B. Desai, J. Fullenkamp, T. Hughes, E. Kennedy, R. McCormick, J. Peters, and D. Zanders. *Hazardous Waste Incineration Engineering* (Park Ridge, NJ: Noyes Data Corporation, 1981).
9. 40 CFR 264.343.
10. 40 CFR 261.24.
11. Kingsbury, G. L., and R. M. Ray. "Reclamation and Development of Contaminated Land: Volume One—U.S. Case Studies," U.S. EPA Hazardous Waste Engineering Research Laboratory, EPA/600/2-86/066 (August 1986).
12. 40 CFR 265.301.
13. Title 22, California Code of Regulations, Section 67281(a).

APPENDICES

APPENDIX A
FEDERAL REGULATORY
APPROACHES

INTRODUCTION

The federal government's approach to regulation of hazardous materials in the United States started out initially with a phase-specific direction. Hazardous contaminants entering the environment were regulated according to whether they were being emitted to the air or to water, were utilized in food, or were involved with constituencies such as the agricultural industry. This approach worked for several decades, particularly following World War II. Nevertheless, by the late 1970s it became apparent that regulations that were "phase-specific" were not accomplishing the objective of protecting public health, safety, and the environment. The phase with the least stringent regulation became the ultimate receptor of pollutants.

As federal regulations developed, a distinction arose between the larger concern for hazardous materials and regulations that dealt with just hazardous waste. In addition, the responsibility for regulating and enforcing hazardous materials laws became split between three or four major federal regulatory agencies.

Figure A.1 illustrates the variety of federal acts that currently regulate hazardous materials in some manner. They also indicate which federal offices or departments are responsible for regulations and enforcement, and finally, those sections of the *Code of Federal Regulations* (CFR) that specifically detail the requirements and where they are to be found.

As can be seen, each agency has its own specific title in the *Code of*

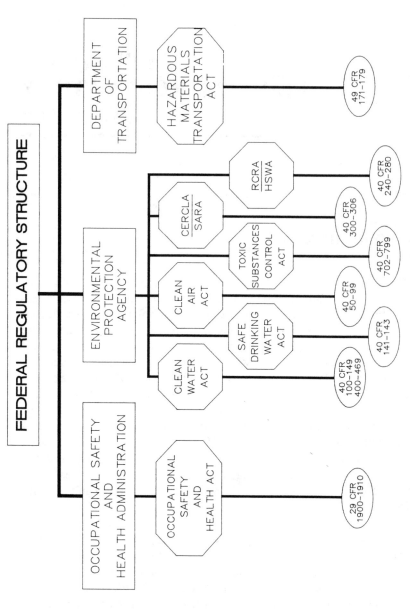

Figure A.1. Major federal agencies and acts of Congress governing hazardous materials.

Federal Regulations. EPA has the 40-CFR series, OSHA has the 29-CFR series, and the Department of Transportation (DOT) has the 49-CFR series.

Today there is a great deal of overlap in the fields that are regulated (such as transportation or treatment), the lists that refer to hazardous material and hazardous waste, and finally, definitions of hazardous material and how it is used.

This appendix will provide an overview of some of the major federal hazardous materials management regulations and will provide an in-depth review of two of the major regulations as amended to date: the Resource Conservation and Recovery Act (RCRA)[1] and the Comprehensive Environmental Response, Compensation, and Liability Act (CERCLA) or Superfund,[2] as it is called. In addition, we will look at the agencies involved and some of the lesser-known "players" in the hazardous materials management regulatory framework.

OVERVIEW OF FEDERAL REGULATIONS

Water

The Clean Water Act[3] and the Safe Drinking Water Act (SDWA),[4] which are both enforced by the EPA Office of Water, deal with two major aspects of water.

Clean Water Act

The Clean Water Act controls the discharge of toxic materials into surface streams. This Act was the outcome of the 1899 Rivers and Harbors Act, which prohibited waterway discharges that could interfere with interstate transportation. In addition, other water pollution control acts were passed between 1948 and 1987.[5]

The earlier acts protected waterbodies by imposing effluent limitations at the source of pollution discharge. The later acts regulated water pollution by defining the level of pollution which the receiving body of water theoretically could handle. Discharge limit provisions and water quality standards were set up. The concept of federal discharge permits was incorporated into the National Pollutant Discharge Elimination System (NPDES).[5]

EPA is required to promulgate toxic discharge requirements for 34 industrial categories covering over 130 toxic pollutants for discharge into surface waters.[5] These toxics include metals, corrosives, and some pesticides. Dischargers of these pollutants are required to use the best available technology (BAT) to achieve these limitations.

Toxic and hazardous waste discharges directly to a receiving body of water are regulated by NPDES permit, whereas discharges acceptable to an industrial or municipal sewer system, such as a publicly owned treatment works (POTW), are allowed without a federal permit.

Safe Drinking Water Act

The Safe Drinking Water Act, passed in 1975, was established to protect groundwater and drinking water sources. This act has set up two levels of drinking water standards to limit the amount of contamination that might be found in drinking water: primary standards with a maximum contaminant level (MCL) to protect human health, and secondary standards that deal with color, taste, smell, or other physical characteristics. This law also required the EPA to establish maximum contaminant level goals (MCLGs) at levels lower than the MCLs for each contaminant which might have an adverse impact on the health of an individual.

The SDWA includes over 83 contaminants for regulation. These include synthetic organic compounds, inorganic chemicals, microbiological contaminants, and radiologic contaminants. Thus, it is more comprehensive than other federal regulations. It also regulates the injection of liquid waste into underground wells used as a means of waste disposal. These regulations assure that this disposal method does not damage the quality of groundwater and groundwater aquifers. Specifically, injection wells are not allowed within one-fourth of a mile of underground sources of drinking water. It is anticipated that this aspect of the SDWA will expand dramatically in the next few years.

Air

The Clean Air Act,[6] which was passed originally in 1963, addresses air pollution that originally had its source in heavily industrialized cities. This act is administered under the EPA Office of Air and Radiation. The Clean Air Act gives the EPA responsibility to set three different standards: National Ambient Air Quality Standards (AAQS), New Source Performance Standards (NSPS), and National Emissions Standards for Hazardous Air Pollutants (NESHAP). These regulations, particularly those that deal with air toxics, are discussed in Chapter 4.

Hazardous Substances

The Toxic Substances Control Act (TSCA)[7] is administered under the EPA Office of Pesticides and Toxic Substances. TSCA was designed as a catch-all regulation to close loopholes perceived to exist in the chemical manufacturing and use laws that were related to environmental protection. It gives EPA broad authority to regulate chemical substances that present a hazard to human health or the environment without regard to specific use, phase, or area of application. This law controls chemicals at the source before they are dispersed into the environment.

The next two acts (RCRA and CERCLA) are now merged under the EPA Office of Solid Waste and Emergency Response (OSWER). These acts are dealt with in later sections.

Worker Protection

The Occupational Safety and Health Act (OSHA)[8] of 1970 was set up as the primary federal law to provide worker protection from toxic substances. OSHA is administered under the Occupational Safety and Health Administration (Department of Labor) and is specifically concerned with setting standards of exposure (mostly air) that are allowed in the industrial workplace. OSHA requires employers to maintain records, track diseases, provide information to employees about the dangers posed by certain toxic substances to which they may be exposed, and set standards for regulating substances that may cause cancer.

Transportation

The Hazardous Materials Transportation Act,[9] which is administered by the DOT, provides a separate management structure for hazardous materials during transport. This level of regulation is appropriate, since there is a higher risk of exposure to health hazards during the transportation of hazardous chemicals than during the manufacture, storage, treatment, or disposal of such substances.

The DOT has the authority to regulate the shipment of hazardous substances that may pose a threat to health, safety, property, or the environment when transported by air, water, rail, or highway. In addition, DOT requires special packaging and routing of hazardous materials. It is interesting to note that the transportation aspect of hazardous materials regulation was originally set up by the federal government in 1865 to protect railroads and railroad workers from explosions that could result from poorly identified packaged explosives and ammunition during the Civil War.

Under the DOT,[5] the Federal Aviation Administration deals with air transport, the Federal Highway Administration deals with highway transport, the Federal Railroad Administration deals with rail transport, and the United States Coast Guard has authority over shipment of hazardous materials by oceangoing vessels.

FEDERAL HAZARDOUS SUBSTANCES ACTS

Resource Conservation and Recovery Act

RCRA and its amendment HSWA (the Hazardous and Solid Waste Act) deal with the ongoing management of solid wastes throughout the country with the emphasis on hazardous waste. Thus, it is keyed to the waste side of hazardous materials, rather than broader issues as they are dealt with by the Hazardous Materials Transportation Act and TSCA (q.v.).[10] It was amended and expanded in 1984 by the Hazardous and Solid Waste Act.

The primary concern of RCRA is to protect groundwater supplies by creating a "cradle-to-grave" management system focusing on treatment, storage, and disposal of hazardous waste. This approach was taken when it was found that improper waste disposal and management practices were polluting the groundwater supplies of the United States. (Approximately 50% of the population relies on groundwater for drinking.) The burden of dealing successfully with hazardous wastes fell on generators and, to a more intensive degree, on those firms dealing with the treatment and disposal of waste materials.

RCRA regulates five specific areas for the management of hazardous waste. These are:

1. identifying what constitutes a hazardous waste and providing classifications for each
2. publishing requirements for generators to identify themselves, which includes notification of hazardous waste activities and standards of operation for generators

3. adopting standards for transporters of hazardous wastes
4. adopting standards for treatment, storage, and disposal facilities (TSDFs)
5. providing for enforcement of standards through a permitting program and legal penalties for noncompliance

The goal of the RCRA program is to regulate all aspects of the management of hazardous waste from the time it is generated until it is finally immobilized, treated, or destroyed.

Waste generation is not prohibited; however, waste minimization is an intrinsic part of current approaches, which include both EPA regulations and private management approaches to eliminating or minimizing waste generation.

Identification

The first of these requirements is for generators to identify their wastes and determine whether the wastes from their operations may be hazardous.

Figure A.2 illustrates the EPA form required by law to be filled out by every generator of hazardous waste. This is a notification to EPA and acts both as a registration form and an application for an EPA Hazardous Waste identification number. This latter number is specific to the waste generator's location and activity and is only good for the specific owner, site, and operation. Completion of this form "gets you into the system" as a generator of hazardous waste.

What constitutes an EPA hazardous waste is outlined in Table A.1. There are two types of hazardous waste under the federal scheme: those that are specifically listed by type of generation or waste stream, and those that possess specific chemical or physical properties (characteristics).

The first type is keyed by a specific letter (F, K, P, U) and refers to the types of waste that are regulated under RCRA regardless of the concentrations of chemical contamination in those waste streams. The second, "characteristic" wastes (D series) may vary in contaminant concentrations or other properties, and thus the waste material may or may not be classified as hazardous.

Please print or type with ELITE type *(12 characters per inch)* in the unshaded areas only

Form Approved OMB No 2050-0028. Expires 9-30-88
GSA No 0246-EPA-OT

United States Environmental Protection Agency
Washington, DC 20460

⬥EPA Notification of Hazardous Waste Activity

Please refer to the *Instructions for Filing Notification* before completing this form. The information requested here is required by law *(Section 3010 of the Resource Conservation and Recovery Act)*.

For Official Use Only

Comments

C
C

Installation's EPA ID Number   Approved   Date Received *(yr.   mo.   day)*

C   T/A C
F   1

I. Name of Installation

II. Installation Mailing Address

Street or P.O. Box

C
3

City or Town   State   ZIP Code

C
4

III. Location of Installation

Street or Route Number

C
5

City or Town   State   ZIP Code

C
6

IV. Installation Contact

Name and Title *(last, first, and job title)*   Phone Number *(area code and number)*

C
2

V. Ownership

A. Name of Installation's Legal Owner   B. Type of Ownership *(enter code)*

C
R

VI. Type of Regulated Waste Activity *(Mark 'X' in the appropriate boxes. Refer to instructions.)*

A. Hazardous Waste Activity	B. Used Oil Fuel Activities
☐ 1a. Generator ☐ 1b. Less than 1,000 kg/mo.	☐ 6. Off-Specification Used Oil Fuel *(enter 'X' and mark appropriate boxes below)*
☐ 2. Transporter	
☐ 3. Treater/Storer/Disposer	☐ a. Generator Marketing to Burner
☐ 4. Underground Injection	☐ b. Other Marketer
☐ 5. Market or Burn Hazardous Waste Fuel *(enter 'X' and mark appropriate boxes below)*	☐ c. Burner
☐ a. Generator Marketing to Burner	☐ 7. Specification Used Oil Fuel Marketer *(or On site Burner)* Who First Claims the Oil Meets the Specification
☐ b. Other Marketer	
☐ c. Burner	

VII. Waste Fuel Burning: Type of Combustion Device *(enter 'X' in all appropriate boxes to indicate type of combustion device(s) in which hazardous waste fuel or off-specification used oil fuel is burned. See instructions for definitions of combustion devices.)*

☐ A. Utility Boiler   ☐ B. Industrial Boiler   ☐ C. Industrial Furnace

VIII. Mode of Transportation *(transporters only — enter 'X' in the appropriate box(es)*

☐ A. Air   ☐ B. Rail   ☐ C. Highway   ☐ D. Water   ☐ E. Other *(specify)*

IX. First or Subsequent Notification

Mark 'X' in the appropriate box to indicate whether this is your installation's first notification of hazardous waste activity or a subsequent notification. If this is not your first notification, enter your installation's EPA ID Number in the space provided below.

C. Installation's EPA ID Number

☐ A. First Notification   ☐ B. Subsequent Notification *(complete item C)*

EPA Form 8700-12 (Rev. 11-85) Previous edition is obsolete   Continue on reverse

Figure A.2(a).   EPA hazardous waste activity form (front).

ID — For Official Use Only												T/A	C
C													
W													1

X. Description of Hazardous Wastes *(continued from front)*

A. Hazardous Wastes from Nonspecific Sources. Enter the four-digit number from 40 CFR Part 261.31 for each listed hazardous waste from nonspecific sources your installation handles. Use additional sheets if necessary.

1	2	3	4	5	6
7	8	9	10	11	12

B. Hazardous Wastes from Specific Sources. Enter the four-digit number from 40 CFR Part 261.32 for each listed hazardous waste from specific sources your installation handles. Use additional sheets if necessary.

13	14	15	16	17	18
19	20	21	22	23	24
25	26	27	28	29	30

C. Commercial Chemical Product Hazardous Wastes. Enter the four-digit number from 40 CFR Part 261.33 for each chemical substance your installation handles which may be a hazardous waste. Use additional sheets if necessary.

31	32	33	34	35	36
37	38	39	40	41	42
43	44	45	46	47	48

D. Listed Infectious Wastes. Enter the four-digit number from 40 CFR Part 261.34 for each hazardous waste from hospitals, veterinary hospitals, or medical and research laboratories your installation handles. Use additional sheets if necessary.

49	50	51	52	53	54

E. Characteristics of Nonlisted Hazardous Wastes. Mark 'X' in the boxes corresponding to the characteristics of nonlisted hazardous wastes your installation handles. *(See 40 CFR Parts 261.21 — 261.24)*

☐ 1. Ignitable *(D001)* ☐ 2. Corrosive *(D002)* ☐ 3. Reactive *(D003)* ☐ 4. Toxic *(D000)*

XI. Certification

I certify under penalty of law that I have personally examined and am familiar with the information submitted in this and all attached documents, and that based on my inquiry of those individuals immediately responsible for obtaining the information, I believe that the submitted information is true, accurate, and complete. I am aware that there are significant penalties for submitting false information, including the possibility of fine and imprisonment.

Signature	Name and Official Title *(type or print)*	Date Signed

EPA Form 8700-12 (Rev. 11-85) Reverse

Figure A.2(b).   EPA hazardous waste activity form (back).

Table A.1 RCRA Hazardous Waste Categories

Listed wastes

"F" List   Nonspecific sources
(e.g., nonhalogenated solvents, certain
process wastes [e.g., electroplating
involving cyanides])

"K" List   Specific sources/
(e.g., wastewater from chrome yellow   Processes
and orange pigment production; heavy
ends and distillation residues from
production of CCl_4)

"P" List   Specific substances
(e.g., cyanogen chloride; dieldrin, carbon
disulfide)

"U" List   Off-spec products
(e.g., benzal chloride, aniline)   and intermediates

Mixture rule: An F, K, P, or U listed waste mixed with solid waste
renders entire mixture hazardous.

Characteristic wastes (chemical or physical properties)

Ignitable (D001):   Flashpoint < 140°F ASTM
Corrosive (D002): $2 \geq pH \geq 12.5$
Reactive (D003):   Reacts violently or generates pressure
Toxic (D004–D017):   Specified concentrations

A mixture of "characteristic waste" and common trash will be
hazardous if it still demonstrates characteristic for listing. The burden
is on the generator to show absence of a characteristic.

The four characteristic properties that qualify a waste as a RCRA
waste material are:

1. ignitability
2. corrosivity
3. reactivity
4. toxicity

Of these four characteristics, ignitability and toxicity bring the largest
quantities of waste materials into the RCRA hazardous waste category.

Ignitability refers to the characteristic of being able to sustain combustion and includes the category of flammability (ability to start fires when heated to temperatures less than 140°F).

Corrosive wastes may destroy containers, contaminate soils and groundwater, or react with other materials to cause toxic gas emissions. Corrosive materials provide a very specific hazard to human tissue and aquatic life where the pH levels are extreme.

Reactive wastes may be unstable or have a tendency to react, explode, or generate pressure during handling. Also, pressure-sensitive or water-reactive materials are included in this category.

Toxicity is a function of the effect of waste materials that may come into contact with water or air and be leached into the groundwater or dispersed in the environment. The toxic effects that may occur to humans, fish, or wildlife are the principle concerns here.

There is a possibility for delisting a listed hazardous waste stream by petition to EPA. These procedures are appropriate for a listed waste that can be shown to be nonhazardous. However, there are a number of legal requirements regarding what must be in the petition, and the generator must submit evidence that such a waste stream is not hazardous. In addition, formal public notice and hearings are required, and comments on the petition must appear in the *Federal Register* for each specific case.

It should be noted that the burden of proving that a waste stream is nonhazardous always falls on the generator of the waste.

In addition to these concerns, the generator must also determine if the waste stream is mixed with other materials that would qualify it as a hazardous waste. Thus, it is a good practice for a generator of hazardous waste to keep tight control of hazardous materials rather than mixing them with common trash prior to disposal.

Standards for Generators

A generator is any "person" whose act or process produces hazardous wastes, or whose act first causes hazardous waste to become subject to RCRA regulations. Thus, a definition of a generator may refer to a site as well as the waste material. The requirements for a generator are outlined below.

It is the responsibility of the generator to identify hazardous waste in its form and content, as well as obtaining a site-specific EPA identification number. This number identifies the generator and is used to track the waste from a specific site throughout its entire history, from "cradle to grave." This also provides the EPA with a mechanism to solve the problem of the midnight dumper. It should be noted that generators may only let their wastes be transported by haulers who themselves have a unique EPA identification number.

One provision allowed to the generator is that he or she may temporarily store waste for a period of 90 days prior to shipment of that waste offsite, even if the facility is not a TSDF. Such storage, however, can only be in tanks and containers and may not be on land, in fields, or in waste piles.

Manifests

Once the waste is ready for shipment offsite to a TSDF, a manifest must be completely filled out that gives explicit information as to the type, quantity, amounts, and chemical concentrations of the waste. It also identifies, through every step of the trip, exactly who is handling the waste and who is receiving it (see Figure 6.2).

The manifest itself is a multisheet printed form of which at least one copy goes to the disposal facility, another copy goes to the state or federal agency, and another copy is ultimately returned to the generator to complete his records.

The manifest concept is central to the entire RCRA hazardous waste management scheme. It is the legal document that tracks the material and indicates what its ultimate disposition will be. In addition, EPA ID numbers are required for every authorized transporter and TSDF that will be handling that material. It should be noted that upon signing that manifest, the generator is also committing to implementation of a waste minimization program.

In addition, RCRA requires that the generator be totally responsible for all preparations, shipping containers, packaging, labeling, marking, and placarding of the hazardous waste to be handled by the transporter. Each container must also be specifically marked with the generator's name and address, manifest number, and the date at which the accumulation of material in that waste container began.

Finally, generators are required to keep and maintain three types of records for a minimum of three years from the date of acceptance of the waste by the transporter:

- copies of the signed waste manifest, which have been returned from the TSDF

• copies of reports filed with the EPA (including biennial waste minimization reports and exception reports, which deal with errors or omissions in waste manifests)
• records of waste results, waste characterization analyses, or other tests showing that the waste is hazardous

Transporter Standards

Transportation requirements for hazardous materials are covered in depth in Chapter 6. The key legal requirements for transporters are:

1. No permit is required for a hauler of hazardous materials; but the hauler must have an EPA ID number.
2. Identification of the source and notice of delivery of a shipment of waste to a TSDF is required.
3. Proper labeling of the transported waste is required.
4. The transporter must comply with all aspects of the manifest system, including his or her own EPA number and signature on the manifest.
5. The waste must be delivered only to the designated TSD facility named on the manifest (the facility must also hold a RCRA permit).
6. A three-year retention period is required for each manifest handled by the hauler.

In the event of an accidental or intentional release of the hazardous waste during transportation, the transporter is responsible for cleanup.

Standards for Treatment, Storage, and Disposal Facilities

The burden of RCRA falls most heavily upon TSDFs. The regulations are very specific, very detailed, and very costly for every step of the process. The heart of the standards for TSDFs are found in the permit requirements.

There are some facilities or activities that are exempt from a permit requirement.[11] These are:

• facilities regulated by other acts (such as the Safe Drinking Water Act [underground injection] or the Marine Protection Research and Sanctuaries Act [ocean dumping]), or POTWs

- facilities that have a previous permit or are purely recycling operations
- haulers, farmers, temporary storage operations (90 days or less), or wastewater treatment units discharging to a POTW;
- the addition of absorbent material to hazardous waste in a container

All other TSD operations must operate under an RCRA permit.

Part A of the EPA permit application gives the location and ownership. Part B of the permit application contains specific details of operation, responsibility, and plans for the long-term operation of any TSDF. The permit itself is specific to the waste stream(s) and has a limited life span of 10 years. This is subject to renewal under appropriate circumstances. The information that must be provided in each of the sections of the Part B permit application is indicated below.

Waste analysis. Details on the specific physical and chemical characteristics of each of the hazardous wastes coming to the facility must be provided in the Part B permit application. Representative sampling and analysis must be repeated on shipments as often as necessary to ensure that the waste analysis plans are accurate and up to date. Procedures to be followed by the operator of the facility must be set forth in this section of the appliction.

Security. Prevention of entry of unauthorized persons or livestock to the TSDF must be detailed in this section of the permit application. This involves a security system including barriers, surveillance, controlled entry, and posted danger signs.

Inspections. An inspection plan specifically tailored to the activities at a TSDF is required to be prepared and implemented. This includes inspections of facilities that are expected to show deterioration as time goes on, or areas such as loading docks where spills or other incidents may occur. These inspections are performed by the owner and operator of the facility; therefore, inspection reports must be made and kept on file at the facility.

Personnel training. TSDF personnel are required to have experience and training in the areas to which they are assigned. Instruction may be provided either in a classroom or on the job, but it must be directed by a person who is trained in hazardous waste management practices. The details of such a training program must also be included in the permit application. These requirements are in addition to regulations requiring

OSHA training for all personnel handling hazardous waste (29 CFR 1910.120).[12]

Special handling. Procedures for handling waste that may be reactive, ignitable, or incompatible must also be detailed. Typically, this includes procedures for operator waste handling, posting of signs, and detailed instructions as to segregation or placement of waste materials according to their characteristics. This prevents incompatible wastes that may generate specific health or environmental hazards from being commingled at the facility. Documentation of these procedures is also required.

Location standards. Location of a new TSDF within 200 feet of a seismic fault, or in a 100-year floodplain, is prohibited. For existing facilities, potential mitigation measures for the latter case are required.

Preparedness and prevention. Detailed plans to eliminate the possibility and effect of fires, explosions, or spills at the TSDF are also required in the permit application. Preparatory measures include such things as alarms and communication systems, systems for emergency help, installed fire control equipment, and decontamination equipment on site. Plans for cooperation with local emergency agencies are also to be detailed.

Contingency plan and emergency procedures. Contingency plans for the facility to minimize hazards to human health and environment due to explosion, fires, or releases of hazardous waste are required. The contingency plan is a preformulated set of responses to any given emergency, and as such, will name emergency coordinators and list emergency equipment that will be available. Recently, the requirements regarding maintenance of contingency plans were extended to apply to any facility that generates, handles, or stores hazardous materials onsite. They are more closely evaluated in Chapter 8.

Manifest systems, recordkeeping, and reporting. Complete records on how all waste shipments received at the facility will be handled must be detailed in the Part B permit application, particularly with respect to discrepancies that may be noted.

The original manifest that arrives with the waste must be retained by the TSDF for a period of three years. Operating records to show how the waste was handled must also be maintained at the facility and available for inspection.

Other operating records, such as waste analyses, testing or analytical results, and other facility closure estimates are also included in the operating records that must be maintained.

In addition, a biennial report is to be filed in March of every even-numbered year covering all hazardous waste management activities during the preceding year.

Closure and postclosure care. This section of the Part B permit application details specifically how the TSDF will be closed, how the equipment will be decontaminated, and how any hazardous waste remaining will be removed, disposed of, or manifested offsite.

Postclosure details must be included to determine in advance the specific activities that the operator will perform with respect to monitoring of environmental hazards and maintenance activities at the facility for 30 years after closure.

Certain facilities, such as incinerators that do not pose a long-term environmental risk after closure, would not be required to give extensive details on postclosure requirements.

Financial responsibilities. This section details how the facility will provide funding for the closure and postclosure activities detailed in the earlier section, in order to avoid CERCLA-type activities (cleanups) at a future date for an existing TSDF. These details must spell out the cost of closing the facility and the annual cost of the postclosure monitoring and maintenance program required under RCRA.

Land Management Units

Each specific type of TSDF has its own facility operating standards, including standards for containers, for tanks, and for all types of land management units such as surface impoundments, waste piles, and landfills. Landfills receive the greatest degree of scrutiny under these regulations, which specifically require selective segregation of waste in the different cells of the landfill and provisions for protection of surface water and groundwater from leachate and surface runoff.

In addition, the specific types of liners necessary to mitigate against any potential liquid migration are also required. Current law requires a double-leachate, double-liner system on top of an impermeable natural liner, such as clay. An impermeable cap is also required at closure of a landfill.

Groundwater protection plans for each type of land management unit are specifically required. This means monitoring existing groundwater quality, before, during, and after facility operations, by laying out and installing monitoring wells and establishing procedures for specific sampling and analysis plans, detection monitoring and compliance monitoring programs, and finally, a corrective action program if waste materials are found to be migrating from the facility. The corrective action program is similar to CERCLA cleanups and is being expanded.

Incinerators

For incineration units there are two major requirements: first, the preparation of a trial burn plan, and second, specific performance standards for air emissions. The trial burn plan consists of information on the waste feed, the performance of the incinerator, the operating temperature, a list of waste materials to be burned, and the period during which the incinerator will be operating.

The principal organic hazardous constituents (POHCs) must also be identified in the waste streams. Specific information must be spelled out on testing of the incoming waste, the stack air emissions, water emissions, and ash content of hazardous wastes to be burned.

The performance standards consist of three specific items. These include a minimum Destruction and Removal Efficiency (DRE) of 99.99% for each POHC designated for a particular waste feed, a removal efficiency of 99% for hydrochloric acid from the exhaust gas, and finally, a particulate emission standard of 180 mg per dry standard cubic meter corrected to 7% oxygen. These performance standards are in addition to any state or local requirements for a permit to operate.

Chemical, Physical, and Biotreatment Units

While other treatment operations are possible, EPA has not attempted to regulate these treatment processes due to the small number of units that are currently in existence. These, however, are most likely covered under local regulations.

CERCLA (Superfund)

The mission of CERCLA (Superfund) is to clean up hazardous waste disposal mistakes of the past and to cope with emergencies of the present. In some cases, Superfund is dealing with problems whose origins go back to the turn of the century. The program is an offshoot of the Clean Water Act of 1972 that provided for the formulation of a National Contingency Plan (NCP) dealing with emergencies involving hazardous waste. These plans have undergone many refinements, but they are still the guiding principles behind the implementation of CERCLA.

It should be noted that CERCLA is a combination of preexisting concepts. The Spill Prevention Provision (Section 311 of the Clean Water Act),[13] is borrowed from RCRA. Also, Superfund is primarily a liability law, not a regulatory law. As such, it is concerned primarily with funding of cleanups, whether from spills or past activities.

CERCLA's foundation is that the generator of a waste is liable for the cost of cleaning up the consequences of that waste, regardless of how legal a waste disposal practice may have been in the past. It is the "doctrine of continuing harm" which states that, while the waste disposal may have been legal in the past, the current impact on the environment is the threat at issue. Therefore, the generator (or "owner") of the waste causing the problem is still financially liable.

There are four key concepts in CERCLA: reporting, remedial action, funding and emergency planning, and community right-to-know. These are outlined in the following sections.

Reporting

CERCLA imposes on the President a requirement that the experience of the program be reported to Congress at least every four years. This report evaluates such things as the liability standards, performance of the taxing and funding programs, and all other aspects of the law. Also included are the feasibility of private insurance plans, siting and location of TSDFs, and the adequacy of existing statutory and legal remedies for harm caused by releases to the environment.

Remedial Action

CERCLA has set up a scheme for analyzing the impact of contaminated sites under a hazard ranking system. From this hazard ranking system, a list of prioritized disposal and contaminated sites is compiled. This list becomes the National Priorities List (NPL) when promulgated, and sites on it receive priority in cleanup attention from the federal government under Superfund.

Once a site is listed on the NPL, those generators who contributed waste to that site become potentially responsible parties (PRPs) who are therefore liable for paying for the activities associated with cleaning up the site. The option is given to the responsible parties to do the cleanup; however, if they default, the federal government is required to step in, perform the activities in the remedial process, and then bill all those who have the financial resources to pay. Thus, the concept of the "deep pocket" arose in paying for cleanup of such contaminated sites.

The remedial process itself requires a number of separate activities to be performed under the law. The NCP details how the President and the EPA are to go about analyzing each of these sites and the procedures they must follow in such analysis, as well as how they reach their decisions.

First come the hazard rankings and studies whereby a particular site becomes listed. This is followed by the Remedial Investigation/Feasibility Study (RI/FS) process in which analytical data and characteristics of the site are evaluated, and feasibility studies are performed to determine what measures could potentially be used to clean up that site. This is followed by an extensive public participation and comment period, and analysis of all the technical details that went into that feasibility study.

The final step is a written Record of Decision (ROD), a legally binding document in which the EPA states what alternative will be chosen to "clean up" that particular site.

Following the signing of the ROD, the design and construction of remedial measures are implemented.

Key changes to CERCLA under the Superfund Amendments and Reauthorization Act of 1986 (SARA)[14] require that other applicable, relevant, and appropriate federal, state, and local regulations be met in the chosen cleanup alternative.

It is noteworthy that under the remedial process, permits are not required from any agency of federal, state, or local governments for a cleanup activity provided that the hazardous materials do not leave that site.

Also under SARA, the congressional directive is that alternatives providing for permanent solutions are to be preferred over those that merely contain hazardous waste at the site or attempt to stop migration of hazardous materials.

Funding

The PRPs are generators (or owners) who contributed waste to the site. They are charged with funding the activities under the remedial process. However, at those sites where there are no PRPs available, the Superfund will step in and finance those activities. This money is collected from national taxes on petroleum and listed chemicals, an environmental tax on corporations with annual revenues exceeding $2 million, a tax on hazardous waste receipts at TSDFs, and general revenue appropriations from the federal treasury.

For all finalized Superfund cleanup decisions made, states are required to contribute at least 10% of the total cost for orphan facilities (those without known responsible parties), provided they agree with the chosen remedial action.

Emergency Planning and Community Right-to-Know

Title III of SARA implemented a new free-standing act. This "Community Right-to-Know" act requires federal, state, and local governments and industry to work together in developing emergency plans and reporting on hazardous chemicals. These requirements are built on EPA's chemical emergency preparedness program and numerous state and local programs aimed at helping communities deal with potential chemical emergencies. The Community Right-to-Know provisions allow the public to obtain information about the presence of hazardous chemicals in their communities and releases of these chemicals into the environment.

Title III has four major sections. These deal with emergency planning, emergency notification, Community Right-to-Know reporting requirements, and toxic chemical release reporting.

Toxic Substances Control Act (TSCA)

TSCA is designed as a gap-filling law for all other acts that attempt to manage hazardous materials, either by process or by phase. Under TSCA, EPA defers to other agencies or other acts for enforcement action if those agencies have statutory authority with a certain identified problem.

TSCA specifically requires that chemical manufacturers and importers provide EPA with a Pre-Manufacture Notice (PMN) and provide available health and environmental effects data at least 90 days prior to manufacture or sale of any chemical or (nonfood) chemical additive.

The law provides a risk-benefit provision similar to other federal legislation. Under TSCA, EPA is required to consider the benefits of a substance to economic and social well-being, the risks posed by alternative substances, and the possible health and economic problems that could result from regulation of a substance. The law bans the manufacture of polychlorinated biphenyls (PCBs). Asbestos is also specifically regulated by TSCA.

Occupational Safety and Health Act

OSHA has been established to provide standards of allowable exposure to toxic chemicals in the workplace for persons who could potentially be exposed to a hazardous substance for a period of 40 hours per week. Specifically, these standards were based on exposures set for a healthy adult male weighing 70 kilograms. OSHA has established standards for over 20 toxic or hazardous substances and over 400 toxic air contaminants. In setting these standards, OSHA evaluates acute, chronic, and carcinogenic health effects. OSHA standards include setting of the Permissible Exposure Limit (PEL), labeling standards for equipment, standards for personal protection, engineering control procedures to mitigate exposure, and monitoring requirements for the health of workers.

Federal Food, Drug, and Cosmetic Act (FFDCA)

The FFDCA[15] was passed in 1938 and evolved from federal involvement in drug protection regulations from the 1840s. The Food and Drug Administration (FDA) is the regulatory agency to determine the safety of foods, drugs, medical devices, and cosmetics. Major provisions of this law ban the intentional addition to food of substances known to cause cancer in animals (the Delaney Amendment),[16] prohibit pesticide residues on raw agricultural products, and require a pre-use assessment before FDA approval of food additives. The FDA under the FFDCA also sets pesticide residue limits for processed foods.

Federal Insecticide, Fungicide, and Rodenticide Act (FIFRA)

FIFRA[5] establishes a regulatory program for EPA to control the manufacture and use of pesticides. The intent of this law is to prevent adverse effects on the environment and public health. Specifically, the environmental risks posed by new pesticides and their persistent characteristics are involved. Under FIFRA, manufacturers must register all new pesticides with the EPA, and those that were manufactured before 1972 must be reexamined to ensure that they meet current safety standards.

Under this act, EPA must evaluate not only environmental effects, but also economic, social, and health impacts. In addition, EPA may refuse to register pesticides judged unduly hazardous or may impose use restrictions on such pesticides. Restrictions must be printed on the label and enforcement actions can be taken based upon labeling problems. EPA can condition the registration for general use or use with restrictions or may cancel registration of the pesticide. Example of cancelled registration chemicals include DDT, kepone, and ethylene dibromide (EDB).

REFERENCES

1. Public Law 98–616, Hazardous and Solid Waste Act, amendments to RCRA PL 94–580 (42 USC 6901) (1984).
2. Public Law 99–499, Superfund Amendments and Reauthorization Act (1986) (amended CERCLA, 42 USC 9601, 1980).
3. Public Law 92–500, Clean Water Act (1972).
4. 42 USC Section 300f et seq.
5. Arbuckle, J. G., M. A. Brown, N. S. Bryson, G. W. Frick, R. M. Hall, Jr., J. G. Miller, M. L. Miller, T. F. P. Sullivan, T. A. Vander-ver, Jr., and L. N. Wegman. *Environmental Law Handbook*, 8th ed. (Rockville, MD: Government Institutes, Inc., 1985).
6. Public Law 91–604, Clean Air Act (1963).
7. Public Law 94–469, Toxic Substances Control Act (1976).
8. Public Law 91–596, Occupational Safety and Health Act (1970).
9. Public Law 93–633, Hazardous Materials Transportation Act (1975).
10. Fortuna, R. C., and D. J. Lennett. *Hazardous Waste Regulation— The New Era* (New York: McGraw-Hill Book Company, 1987).
11. 40 CFR 264.1.

12. "Occupational Safety and Health Guidance Manual for Hazardous Waste Site Activities," NIOSH/OSHA/USCG/USEPA, NIOSH Publication No. 85-115 (October 1985).
13. 40 CFR 264.52b.
14. Public Law 99-499, Superfund Amendments and Reauthorization Act (1986).
15. 21 USC, Section 346a et seq.
16. 21 USC, Section 348(c)(3)(A).

APPENDIX B
CALIFORNIA REGULATORY
APPROACHES

INTRODUCTION

The regulation of hazardous materials in a state is not a mirror image of federal activity, since the state has entirely independent authority. The agencies enforcing toxic and hazardous materials laws in a state may, however, operate by two mechanisms. They may operate by delegation when they receive authorization from federal agencies such as the EPA. They may also operate and regulate beyond federal authority under the provisions of state law. Indeed, they may be much more stringent than federal authority and may cover areas that the federal government has not acted upon. The exception is for those regulations that could be construed as a restriction on interstate commerce. There are, therefore, separate and distinct legal systems that one must be aware of in the field of hazardous materials management.

States typically operate and regulate under a variety of laws and a corresponding variety of enforcement agencies. Typical of such agencies in California are the California Highway Patrol (CHP), the Department of Health Services (DHS), the local fire departments, the local Air Quality Management Districts (AQMDs), the county Health Departments, the state and regional water quality control boards, and the municipal codes of cities. There appears to be a much higher degree of cooperation between state and local agencies enforcing hazardous materials laws than is typically found between states and the federal government.

However, between regulatory agencies, definitions may differ as well as regulations for hazardous materials management. In addition, administration and enforcement activities corresponding to specific acts of the legislature or regulations of an agency are separate and distinct.

In this chapter, we will look at acts of the California legislature, briefly review some explanations of the state and local agencies, walk through a flowchart for hazardous waste management under state law in California, and briefly review the state's underground tank regulations.

REGULATIONS

The Hazardous Waste Control Act of 1987[1] is the major act dealing with all aspects of hazardous waste management in California.

Table B.1 is a listing of the major articles in that act and the specific topics with which they deal. As can be seen, the topics to some degree follow those of the federal regulations. However, there are certain aspects that are important to be discerned in each of those specific articles. Article 2 gives very specific definitions to terms not found in federal law; i.e., extremely hazardous waste, border zone property, retro-grade material, restricted hazardous waste, and waste. Each of these definitions introduces concepts that must be understood in order to understand the regulations.

Table B.1 California Hazardous Waste Control Act

Article	Subject
2	Definitions
3	Technical advisory committee
3.5	Hazardous waste management plans
4	Listings
4.5	State regulations of existing facilities
5	Standards
5.5	Coordination with federal acts
6	Transportation
6.5	Hazardous waste haulers
7	Treatment, recycling, and disposal technology
7.7	Hazardous Waste Management Act of 1986
8	Enforcement
8.5	Hazardous waste testing laboratories
9	Permitting
11	Hazardous waste disposal land use
12	Financial responsibility: closure and maintenance of facilities

Article 3.5 requires local governments (counties) to prepare waste management plans for waste disposal within each county and for all waste generated in each county in California. This mandate from the state serves to require that each county make provisions for acceptable treatment and disposal facilities for the wastes generated within that county. Key provisions of the law require that treatment and disposal facilities be available as a function of local government and that fees be established for the waste generated in each county.

From these local plans, the state provides a statewide management plan that serves as a comprehensive planning document for the state and is used as an information source for the public and other areas of government.

Two key outcomes of this planning activity, to be provided over the next five years, are (1) an inventory of existing and planned facilities that handle, store, recycle, dispose of, or otherwise manage hazardous waste, and (2) an assessment of the need for additional waste treatment facilities.

Article 4 provides for the listings and the criteria by which wastes are to be classified or exempted from regulation under this act.

Article 4.5 prohibits a city or county from adopting laws which will prohibit the licensing or permitting of a hazardous waste facility. Thus, the state recognizes the need for ongoing hazardous waste treatment and disposal operations.

Article 5 sets the standards that are to be followed by anyone attempting to handle, process, use, store, or dispose of hazardous waste. A variety of state or local agencies that have authority in this field are specifically cited as being required to provide inspection and technical assistance.

Article 5.5[2] requires that the state conform with all federal laws dealing with hazardous materials, including both RCRA and CERCLA.

Transportation requirements are found under Article 6, which specifies that the CHP will be the lead agency, that manifests shall be utilized for all transportation, and that registration for haulers will be through the CHP. Also, a listing of all registered haulers shall be provided by the CHP.

Under Article 7, the DHS is charged with providing studies and technical references for those interested in hazardous waste management, establishing a toxic substances hotline, and providing for a variety of other enforcement, inspection, and surveillance activities dealing with hazardous waste. Notable among these is the requirement that the DHS provide an annual California Hazardous Waste Management symposium. Funds to support the toxic substance enforcement program are specified in this section. The waste receipt rates are charged by the State Board of Equalization based on wastes disposed of at TSDFs in California.

The Hazardous Waste Management Act of 1986[3] is incorporated by reference in Article 7.7. In this section, references to Title 22 of the California Administrative Code (CAC)[4] are made. The act also provides for the ban on certain types of disposal activities.

Table B.2 details the hazardous waste land disposal bans currently in effect, both for California and the federal government. One will note that California is ahead of the federal schedule on banning liquid waste disposal practices at land disposal facilities.

Enforcement procedures and regulations regarding hazardous waste testing laboratories are provided for in Articles 8 and 8.5.

Procedures for permitting hazardous waste management facilities are detailed in Article 9. Under this section, interim status is allowed for existing facilities; it also sets up specific standards for permits by the state, in the event of final authorization of California by the EPA to enforce the provisions of RCRA. Specifications and requirements are very similar to those under RCRA and are included in Article 9.

Provisions for hazardous waste disposal land utilization (border zone property classification) and financial responsibility requirements for closure and maintenance of hazardous waste facilities are provided for in Articles 11 and 12.

While the DHS is the major lead agency with respect to regulation of hazardous waste and hazardous materials in California, a number of other agencies are involved and their regulations are spread through a number of California regulatory codes.

In Table B.3 are three of the codes that deal specifically with hazardous wastes. The CAC includes industrial relations, motor vehicles, public safety, water quality, and DHS. The California Health and Safety Codes deal with hazardous waste control and hazardous waste haulers. The California Motor Vehicle Code deals with transportation of hazardous materials and also with spills and spill abatement.

Porter-Cologne Water Quality Act

The Porter-Cologne Act[5] is set up under Title 23 of the CAC. Regulations dealing with all aspects of surface and subsurface water with respect to hazardous wastes are adopted by the State Water Resources Control Board (SWRCB). Specifically under this act and operating through the Regional Water Quality Control Boards are regulations cov-

Table B.2 Hazardous Waste Land Disposal Bans

Ban Name	Regulation	Date	Discussion
Containerized Liquids	1,2	1/1/83	Bans liquid in containers from landfills; allows use of absorbents to soak up liquids
California List	2	6/83 to 1/85	Bans disposal of liquids containing certain metals, low pH, cyanides, PCB's and halogenated sludges
Bulk Liquid in Landfill	1	5/8/85	Bans all hazardous liquids in landfills; liquids still can be placed in surface impoundments
Absorbent Ban	1	7/11/86	Issuance of guidance document banning use of sorbents for bulk disposal
CA. Liquid Ban	1	10/1/86	Bans disposal of liquids in landfills
Solvent Ban	1	11/8/86	Bans disposal of liquid solvents
"California List"	1	7/8/86	Federal adoption of an existing California ban
Lab Pack Ban	2	7/8/87	Lab packs are small containers within 55-gallon drums surrounded by absorbent
Heating Value Ban	2	1/8/88	Bans disposal of hazardous waste with a heating value greater than 3000 Btu/lb. unless treated
First RCRA Ban	1	8/8/88	Bans first round of RCRA listed waste
Solvent Soil Ban	1	11/8/88	Bans soil from CERCLA cleanup sites contaminated with solvents
Second RCRA Ban	1	6/8/89	Bans second round of RCRA wastes
California Disposal and VOC Ban	2	1/1/90	Bans disposal of all nontreated hazardous wastes and those with a high VOC content
Final RCRA Ban	1	5/8/90	Bans third round of RCRA listed wastes and all RCRA characteristic wastes

1: Federal.
2: California.

Table B.3 California State Codes Governing Hazardous Wastes

California Administrative Code*

Title 8	Industrial relations
Part 1	Department of Industrial Relations
Chapter 4	Division of Industrial Relations
Subchapter 7	General Industrial Safety Orders
Group 16	Control of hazardous substances
Title 13	Motor vehicles
Chapter 2	Department of California Highway Patrol
Subchapter 6	Hazardous material
Title 19	Public safety
Chapter 1	State Fire Marshall
Subchapter 11	Transportation of flammable liquid tanks on highways
Title 22	Social Security
Part 2	Health and Welfare Agency, Department of Health Services Regulations
Division 4	Environmental health
Chapter 30	Minimum standards for management of hazardous and extremely hazardous wastes
Title 23	Water quality
Chapter 3	Hazardous material disposal
Subchapter 15	Discharge waste to land
Article 2	Waste classification and management

California Health and Safety Code

Division 20	Miscellaneous health & safety provisions
Chapter 6.5	Hazardous waste control
Article 6.5	Hazardous waste haulers

California Vehicle Code

Division 2	Administration
Chapter 2	Department of California Highway Patrol
Article 4	Highway spill containment and abatement of hazardous substances
Chapter 2.5	Licenses issued by California Highway Patrol
Article 4	Transportation of hazardous waste
Article 6	Hazardous waste inspection
Division 14.1	Transportation of hazardous material
Division 14.8	Safety regulations
Division 15	Size, weight, and load

*Titled Code of California Regulations (CCR) – 1988.

ering the aspects of spills, surface impoundments, underground storage tanks, and regulation of pretreatment and POTWs, as well as landfills.

Well water permits, waste discharge permits, and NPDES permits are all administered under Title 23 of the CAC.

Solid Waste Control Act

Title 14 of the CAC[6] deals specifically with landfills, county solid waste management plans, and waste transfer stations. Under this act, the California Waste Management Board operates in conjunction with county health departments in providing protection to the environment from contamination by land disposal facilities.

Land contamination as well as potential contamination of air and water may be regulated indirectly under the waste management board within the jurisdictions of landfills and solid waste.

Safe Drinking Water Act

The passage of the Safe Drinking Water and Toxic Control Enforcement Act of 1986 (Proposition 65)[7] changed the direction of regulation of drinking water supplies.

Under this act, the governor is specifically named as being the party responsible for developing a list of chemical carcinogens and teratogens that may pose a threat to individuals through either cancer or reproductive toxicity. In addition, the act requires exposure notifications to persons exposed to these chemicals.

The act sets up a prohibition for discharge of these materials to any potential drinking water source, and provides for increased criminal penalties, citizens' lawsuits, and notification requirements for local agency inspectors. Because these new requirements result in more stringent standards than under previous law, the practical effect of these requirements would be the imposition of new conditions for issuance of permits for discharges into potential sources of drinking water. In order to implement the new regulations, state agencies responsible for issuing permits are required to alter their regulations.

The full impact of this law is unknown at the time of this writing. One key item is that the discharge prohibition includes the phrase "significant amount," which is defined as being "below limits of detection" by analytical chemical methods. Such a designation imposes drastically lowered contaminant concentrations for persons discharging either carcinogens or chemicals causing reproductive toxicity.

Other Enforcing Acts and Agencies

The Mulford-Carrell Act[8] provides the authority to the California Air Resources Board (CARB) and the local air pollution control districts under Title 17 of the CAC to regulate all sources of air emissions in the State of California.

The CHP, under the Vehicle Code,[9] has wide authority with respect to transportation and spills on public highways in California. In addition, the CHP issues licenses to haulers of hazardous materials.

Under Title 19 of the CAC,[10] the State Office of Emergency Services is authorized to work with health and fire departments in providing for emergency response plans, hazardous materials releases, and area business plans.

Figure B.1 shows the interrelationships between the adopted acts, the enforcing agencies at the state level, references to the CAC, and finally the local authorities that jointly administer these programs.

STATE AND LOCAL AGENCIES

A description of each of the major state agencies and their specific area of enforcement authority for hazardous materials is given below.

California Air Resources Board (Titles 13 and 17)

The CARB was established by the legislature in 1967 to establish and coordinate efforts to attain and maintain ambient air quality standards, to conduct research into the causes of, and the solution to, air pollution, and to control emissions from vehicular sources. The board has overall responsibility for the state's efforts to reduce air pollution.

In the area of airborne toxics, the board has both general and specific regulatory authority. Pursuant to its general authority to establish ambient air quality standards, the board may adopt standards for substances that may be considered toxic. Similarly, the board's regulation of vehicle emissions includes fuel regulations restricting the amount of particular toxic substances in vehicle fuel. The local air pollution control districts

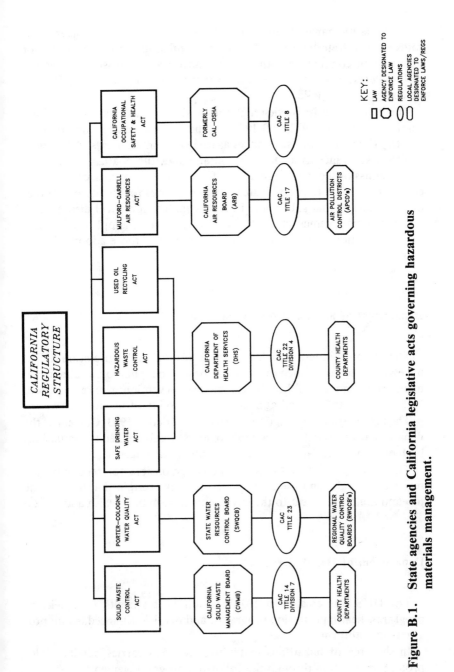

Figure B.1.   State agencies and California legislative acts governing hazardous materials management.

and air quality management districts have jurisdiction over nonvehicular sources. The state board establishes by regulation the test methods for determining compliance with district nonvehicular emission standards, including test methods for substances that have been declared by EPA to be hazardous air pollutants.

The board's specific toxic regulatory authority was established by legislation enacted in 1983 (AB 1807).[11] The statute requires the board to identify (by regulation) substances as toxic air contaminants. After identifying a substance as a toxic air contaminant, the board is directed to adopt emission control regulations for vehicular sources of toxic air contaminants and to promulgate control measures for stationary sources of emissions of that substance. Local air pollution control districts and air quality management districts are then required to adopt control measures for stationary sources at least as stringent as those adopted by the board.

Department of Food and Agriculture (Title 3)[12]

The responsibilities of the Department of Food and Agriculture (DFA) have grown to cover such activities as pesticide regulation and management, plant and animal health, and disease prevention; worker and public health and safety; and agricultural commodity inspection.

The laws pertaining to agriculture were codified in 1933. Currently, the department's jurisdiction involving toxics is closely related to regulations by federal agencies and is shared with other state agencies.

The Division of Pest Management is responsible for regulating the registration, sale, and use of pesticides to make certain that they do the job and are not a threat to the environment or public health and safety.

Department of Health Services

The DHS was created in 1978[12] to administer health care delivery programs, health protection programs, and publicly financed health programs in California.

In the area of hazardous materials, the DHS carries out its health protection function through the following program elements.

Environmental Health Hazard Assessment

The primary objective of the Environmental Health Hazard Assessment Program is to explain the relationships between occupational and environmental exposures to inorganic agents and the subsequent adverse public health effects. The department (1) attempts to distinguish between environmental exposures that pose a health hazard and those that do not; (2) provides technical assistance to local government agencies; (3) summarizes existing health effects information; (4) conducts direct investigations on the human population; (5) produces risk assessment data; and (6) conducts studies pertaining to occupational health research and development.

Toxic Substances Control

The primary objective of the Toxic Substance Control Program is to protect public health and the environment from the effects of toxic wastes. This is accomplished through various activities to regulate hazardous waste generators, TSDFs, and transporters of hazardous waste. Other activities that contribute to the achievement of this objective are (1) hazardous waste facility siting and evaluation; (2) administration of resource recovery and health safety programs; (3) conducting hazardous substance assessment, financial liability, closure, and postclosure maintenance plan reviews; (4) mitigation of sites; (5) coordination of emergency response actions; (6) hazardous waste property evaluation; and (7) assessment of abandoned sites.

Radiologic Health

Under Radiologic Health, the DHS provides protection from the dangers of ionizing radiation and reduces radiation exposure to workers and the public from radioactive materials. The Radiation Standards Unit (1) develops standards and regulations for the training of personnel, design of facilities (in conjunction with the Building Standards Commission), and operations involving the use or redirection of radioactive materials; and (2) develops and enforces standards for the disposal and transportation of radioactive wastes and materials. The Radiation Management Unit registers and licenses users of radiation sources. Inspections and surveys of facilities are conducted to assure that appropriate health and safety standards are followed.

Local Environmental Health

Under Local Environmental Health, the DHS works with local health departments and state institutions to protect an environment from disease and hazards. Local Program Development staff plan and implement local environmental health and toxic substance control enforcement programs. Small Water Systems Surveillance staff provide assistance to county environmental health departments for inspecting small public water systems to obtain compliance with the Safe Drinking Water Act, provide training to county staff responsible for small water system inspection, and coordinate statewide small water system data collection for the EPA.

The DHS is authorized to adopt regulations necessary to administer the above-mentioned program elements. In the regulation development and adoption process, the DHS consults with and receives input from other state agencies (primarily the SWRCB and the CARB), private industry, environmental protection organizations, and the general public.

Department of Highway Patrol (Title 13)

The Department of the California Highway Patrol was formally established by the Legislature in 1929.[12] The passage of this legislation gave statewide authority to the Highway Patrol to enforce laws relating to traffic on county and state highways.

In the area of toxics, the CHP has the authority to adopt regulations governing (1) designated routes and stopping places for explosives transporters; (2) the hazardous materials transportation licensing programs; (3) times and routes for shipments of radioactive materials; (4) flammable and combustible liquid cargo tanks, including their design and construction; and (5) hazardous materials transportation, including the filling, packaging, marking, labeling, and assembly of containers, and the shipper's documentation.

Prior to adopting these regulations, the CHP must consult with local fire prevention officials, transportation companies, explosives manufacturers, and the State Fire Marshal for explosives routes and stopping places; the State Department of Health Services for radioactive materials routes; and the Department's Cargo Tank Advisory Committee for flammable and combustible liquid cargo tanks.

Regulations adopted by the CHP have a direct impact on shippers and carriers of hazardous materials, including explosives and radioactive materials, and on owners of cargo tanks that transport flammable or combustible liquids.

In 1985, legislation was enacted giving the CHP the authority to prohibit transportation of hazardous wastes on selected highways under specified conditions. Additionally, other legislation requires the CHP to coordinate the establishment and approval of routes to be used for the transportation of hazardous materials throughout California. As part of this process, the CHP is required to publish lists of restricted routes and resolve conflicts in routing regulations established by local communities.

Department of Industrial Relations (Title 8)

The Department of Industrial Relations[12] was established by the legislature in 1927 to protect the California work force. Current areas of responsibility include occupational safety and health, primarily through the adoption of General Industry Safety Orders. California OSHA is now only an office under the Department dealing with industrial safety for state and local government workers.

Waste Management Board (Title 14)

The California Waste Management Board[12] was established by the legislature to administer state policy for solid waste management and resource recovery in order to protect the public health and safety and to preserve the environment.

In the area of toxics, the board establishes and administers solid waste handling and disposal standards for facilities that receive both solid and hazardous wastes and for used oil haulers, transfer facility operators, and recyclers.

The board is authorized to adopt regulations governing solid waste handling, storage and disposal, solid waste planning, resource recovery, recycling and litter control, conflicts of interest, financial assurance during facility operation, and the used oil recycling program. The board is not authorized to adopt regulations directly governing toxics.

Prior to adopting regulations on government solid waste handling, storage, and disposal, the board must include public health standards submitted by the DHS and must consider recommendations of the CARB and the SWRCB.

State Water Resources Control Board (Title 23)

The SWRCB[12] was established by the legislature in 1967 to exercise the adjudicatory and regulatory functions of the state in the field of water resources. The board's two major responsibilities are to regulate water quality and to administer water rights. The board's water rights responsibilities involve the issuance of permits and licenses to applicants who desire to appropriate water from streams, rivers, and lakes. The SWRCB carries out its water pollution control responsibilities by establishing wastewater actions against polluters. In addition to the fact that the board's basic water quality powers apply to toxic wastes, the board has been given specific authority to deal with toxic discharges from underground storage tanks, underground injection wells, and surface impoundments.

The SWRCB is authorized to adopt all regulations it deems advisable in carrying out any of its powers and duties. The legislature has specifically directed the board to adopt regulations in several areas dealing with toxics, including waste disposal to land and underground storage tanks.

MANAGING WASTE AT THE LOCAL LEVEL

From the perspective of a generator, it is important to know what steps would be necessary to take a hazardous waste and work one's way through to the safe and legal disposal of that material.

Figure B.2 shows a flowchart for when one is starting out with a waste and attempting to get to the final destination.

Starting at the top, the waste itself has to be determined as hazardous or not. This is typically done by waste characterization analysis. (See Chapter 7.) If one does not have a hazardous waste, the material then can be recycled or disposed of in a safe manner. If the material is hazardous, a further test in California must be performed and that is to determine whether the material is an extremely hazardous waste.

If the waste is extremely hazardous, one is also required to obtain an extremely hazardous waste permit. If the material falls into the normal hazardous waste category, a hazardous waste general license must be obtained from the county, along with an EPA ID number. At that point, one has the option of storing the material for either more than or less than 90 days. In the latter case, one simply stores the material until it is hauled offsite.

If stored for more than 90 days, another application for a permit must

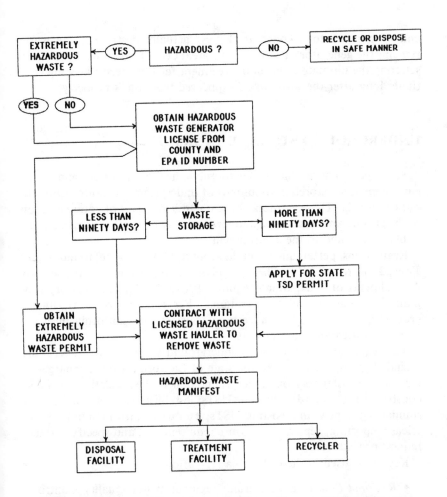

Figure B.2.   Hazardous waste activity flowchart (California).

be made from the state and the EPA as a storage facility (TSDF). Under current state law, the 90 days begin when one has accumulated a minimum of 100 kilograms of hazardous waste on the site.

Once the waste is ready to be removed from the site, one must contract with a state-licensed hazardous waste hauler to remove the waste from the facility. At this point, the uniform hazardous waste manifest must be completed in order to accompany the waste.

The waste then goes to either a permitted disposal or treatment facility or a licensed recycler. While this is the end of the flowchart, it is not the end of a generator's concerns. Unless actually destroyed, the waste remains the generator's forever, and a great deal of care must be taken in selecting the ultimate disposal or treatment facility. Recordkeeping continues long after the waste has disappeared from one's property.

UNDERGROUND STORAGE TANKS

Substances that are hazardous to public health and safety and to the environment are stored in thousands of underground locations throughout California. These substances are potential sources of contamination of the ground and underlying aquifers and may pose other dangers to the public health and to the environment.

Regulations pertaining to underground tanks can be found under Title 23 of the CAC.[13] These regulations are similar to the RCRA Amendments of 1984, which require EPA to develop a regulatory program for underground storage tanks. Per EPA instructions, underground storage tanks that have been registered in California meet the federal notification requirements.

The regulations pertaining to underground storage tanks (USTs) are intended to protect waters of the state of California from discharges of hazardous substances from underground tanks. The regulations establish construction standards for new USTs; establish separate monitoring standards for new and existing USTs; establish uniform standards for release reporting, repair, and closure requirements; and specify variance request procedures.

Key terms used in the California regulations[14] are:

- *Regional Board* — a California Regional Water Quality Control Board
- *Department* — the State Department of Health Services
- *Local Agency* — the department, office, or other agency of a county or city designated pursuant to Section 25283
- *Underground Storage Tank* — any one or a combination of tanks, including pipes connected thereto, which is used for the storage of hazardous substances and which is substantially or totally beneath the surface of the ground
- *Continuous Monitoring* — a system using automatic equipment which routinely performs the required monitoring on a periodic or cyclic basis throughout each day

Key Provisions

Monitoring

Owners of existing underground storage tanks subject to these regulations must implement a visual or alternative monitoring system.[15] This monitoring system must be approved by the local agency having jurisdiction per the Health and Safety Code. The local agency will not issue a permit unless the monitoring system is capable of determining the containment ability of the underground storage tank and detecting any active or future unauthorized releases. The failure to implement an approved monitoring system can result in the local agency requiring closure of the underground storage tank.

The objective of the monitoring program for existing underground storage tanks is to detect unauthorized releases before groundwaters are affected. Groundwater monitoring may be utilized as a primary means of monitoring when the groundwater does not have an actual or potential beneficial use. Minimum monitoring methods are also required.[16]

Standards

Minimum standards have also been established for the construction, installation, and monitoring of new underground storage tanks[17] that contain hazardous substances. New underground storage tanks storing only motor vehicle fuels have different construction standards.

Reporting

All unauthorized releases from the primary or secondary container must be reported to the local board. Certain unauthorized releases to secondary containers must be recorded on the operator's monitoring reports. No other report will be required if the leak detection system in the space between the primary and secondary containers can be reactivated within five hours. This is only applicable to new underground storage tanks. All other unauthorized releases shall be reported within 24 hours after the release has been or should have been detected.[18]

Permits

Specific administrative actions must be accomplished by all underground storage tank owners, local agencies, and the SWRCB relative to issuing permits for underground storage tanks.[19] The information that must be submitted by the underground storage tank owner to the local agency as part of a permit application, conditions local agencies must include in all permits issued and conditions local agencies must meet prior to issuing a permit, and conditions which must be met by underground storage tank owners when requesting "trade secret" designations are also detailed.

WASTE CHARACTERIZATION AND ANALYSIS METHODS

The state of California has its own methods for analyzing hazardous wastes and hazardous materials. These are found under Title 22 of the CAC.

Table B.4 summarizes the basic methods and criteria for analyses in Title 22.

EPA reference test methods in SW846 are referred to as being appropriate for some analytical methods in California. For toxicity criteria, California determines not just the concentration of a substance but its ability to impact living species (i.e., by oral, skin, or inhalation methods), or by the ability of certain waste streams to kill small test fish such as fathead minnows or golden shiners in 96 hours. In addition, California broadens the scope of its methods to include anything that is carcinogenic, toxic, persistent, or bioaccumulative.[20]

California also defines two threshold limit concentrations for use in determining whether a material is hazardous or not. The STLC (soluble threshold limit concentration) sets a lower concentration limit (mg/L) for organic chemicals and elements (in their extracted solutions) that qualify as hazardous materials. The TTLC (total threshold limit concentration) is based upon the wet weight of material and specifies what concentration will qualify a substance as hazardous if found in concentrations greater than listed.[21]

If a concentration of a substance is less than the STLC, there is no further need for testing—it is nonhazardous. If the concentration of

Table B.4 California Title 22 Hazardous Waste Criteria

§66694	• References SW846 (EPA methods)
§66696	• Defines toxicity criteria

           – LD_{50} (oral)   < 5,000 mg/kg
           – LD_{50} (dermal)   < 4,300 mg/kg
           – LC_{50} (inhalation)   < 10,000 ppm
           – 96-hr LC_{50} < 500 mg/Liter
           (fathead minnows/golden shiners fishkill)
           – Acute substances   ≥ 10 mg/kg
           – A hazard to human health or environment (carcinogen; toxic; persistent/bioaccumulative, i.e., DDT)
           – Listed in 40 CFR 261

§66699   • Defines two levels of hazard for specific substances

           – Soluble threshold limit concentration (STLC) in mg/L of extracted solution
           – Total threshold limit concentration (TTLC) in mg/kg – wet weight of material

§66700   • Describes the waste extraction test (WET)

           – Preparation and extraction only

§66702   • Ignitability criteria

§66705   • Reactivity criteria

§66708   • Corrosivity criteria

---

material is greater than the TTLC, the material is hazardous. However, if the analyzed concentration is between these two levels, one must perform the California Waste Extraction Test (WET).

    The WET is a preparation and extraction procedure.[22] In it one takes the sample, extracts it, and then uses the extract for chemical analysis or for performing the 96-hour LC_{50} fishkill test. If the test is passed, the material is considered nonhazardous.

    Three other analytical methods used under California law[23] are outlined in the CAC. These summarize the criteria for determining ignitability, reactivity, and corrosivity in any particular waste stream. Thus, California covers the range of methods included in federal law as well as adding its own methods for other environmental standards such as the ability to kill fish in a waste extract solution.

AIR TOXICS

The primary state agency involved in regulatory oversight of air toxics is the CARB. The major roles of this California agency are to:

• revise, update, and implement the state implementation plan which indicates how the federal AAQSs are to be attained in California

• establish state ambient air quality standards (which are goals for other pollutants)

• demonstrate "reasonable further progress" in attaining ambient air quality standards as required under the Clean Air Act

• adopt model rules for the regional and local air quality districts based upon statewide studies of emissions and sources

• identify toxic air contaminants along with the DHS and the CARB's Scientific Review Panel

• maintain oversight authority for enforcement of local and state Health and Safety Code regulations

It should be noted that the CARB does not have permitting authority for stationary sources of air toxic emissions. This is the domain of the local air quality district and the EPA (at the national level). Thus, CARB functions primarily in an administrative and planning function for the overall management of air quality in California.

Two other state agencies which have an impact on the regulation of air toxics in California are the DHS and the DFA.

The DFA is responsible for evaluating the health effects of pesticides that may be emitted into the ambient air of California and that may be determined to be toxic air contaminants. In addition to evaluating health effects of pesticides, the director of DFA is charged with adopting and setting up regulations designed to adequately prevent, through the application of best practicable control techniques, any endangerment of public health. These techniques may include labeling amendments, applicator training, restrictions on use patterns or locations, changes in application procedures, reclassifications of pesticides as a restricted material, and cancellation of the registration of a pesticide.

The DHS is charged with providing relevant health information to all other state agencies including the CARB and the DFA. Towards that end, DHS activities include taking the exposure data based on modeling efforts or measured ambient concentrations provided by either the CARB or the DFA and performing risk assessments of potentially identified air toxics for human health effects. The risk assessments are performed by DHS toxicologists. They also provide input to the CARB Scientific Review Panel.

The Scientific Review Panel is an advisory body to the CARB and reviews emissions data, ambient test data, consistency of information, and the approach used by the CARB and the DHS in their reports. The panel makes recommendations and changes to draft air toxic reports prepared by the CARB. The latter must then revise the document and issue it.

Air Toxic Identification and Rule Making

Individual chemicals that may be air toxics are evaluated one at a time before they reach regulatory action by the CARB.

The final report to the CARB, after it has been reviewed by the Scientific Review Panel, will include exposures to the air toxic of concern; ambient concentrations found in California (and around the world); the sources of the air toxic, production, usage, and current stationary source emissions; and the impact of current trends in production, usage, and emissions. The final items include persistence in the atmosphere, chemical and physical properties, and ultimate environmental fate. From this report, a regulation may be drafted by the CARB.

The overall regulation of air toxics in California is a two-step process. The first step is a substance identification process and requires approximately 10 months. This is seen visually in Figure B.3. The process requires a number of administrative and procedural steps in order to reach a final decision on whether the air toxic qualifies for regulation.

The second portion of the process in Figure B.3 is the control decision. Approximately eight months after a substance has been identified as needing regulation as an air toxic, a "regulatory needs report" is prepared. A public hearing is held and CARB may adopt an airborne toxic control measure. Local air pollution agencies are then given a "suggested control measure" (SCM) for adoption into their own individual enforcement regulations. This takes approximately six months more, due to local public hearings and regulatory requirements. Substances identified or up for review as air toxics by CARB are identified in Table B.5.

SUBSTANCE IDENTIFICATION PROCESS

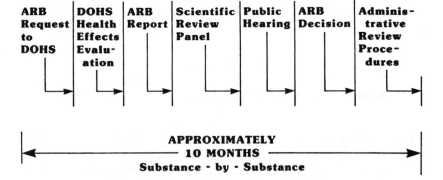

Figure B.3(a). Air toxics identifications process (California Air Resources Board).

Local Air Pollution Control Authorities

Every county in California must have a local air pollution control district or be represented in a regional air quality management district.[24] The South Coast Air Quality Management District (SCAQMD), which consists of the counties of Los Angeles, Orange, and Riverside, along with the nondesert areas of San Bernardino County, will be used as the example of a regional agency. The function and structure of a local county agency is the same throughout California.

Figure B.4 shows the structure of the air quality management district and its role in regulating air toxics and air pollution in general. The governing board of the AQMD functions as a legislative body inasmuch as it adopts regulations under the authority of the state of California. The members are elected officials or public members appointed by elected officials. The SCAQMD personnel provide staff support to the board.

The air quality management district is managed by an executive officer with staff. The local district or regional agency is required to implement

CONTROL
DECISION
PROCESS

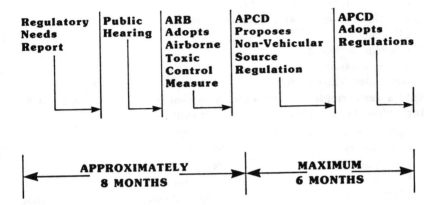

Figure B.3(b). Air toxics control decision process (California Air
Resources Board).

adopted programs and regulations and provide air quality monitoring
services throughout the district. The agency has a comprehensive function inasmuch as it provides engineering services, source testing services,
ambient air monitoring, chemical lab services, and administrative
functions.

The primary goal of the local district, however, is enforcement of the
air pollution control laws. District staff members provide inspection
services and serve as expert witnesses in court cases regarding violations
of air pollution control regulations. They perform ambient air monitoring and planning functions as well as evaluating permits to construct and
permits to operate stationary sources of toxic and criteria pollutant emissions and control equipment. They set general and specific operating
limits for both criteria and noncriteria pollutants to maintain compliance
with the regulations.

The third portion of the local agency is the AQMD Hearing Board.
This is a quasi-judicial body set up by state law,[25] which is independent of
the district board and the district staff. It functions in the role of oversee-

Table B.5 Status of Toxic Air Pollutant Identifications

Substances identified as toxic air pollutants

Asbestos
Benzene
Cadmium
Carbon tetrachloride
Chlorinated dioxins and dibenzofurans
Ethylene dibromide
Ethylene dichloride
Ethylene oxide
Hexavalent chromium

Substances currently under review, scheduled for review, or nominated for review for identification as toxic air contaminants

Substances already in the review process:

Chloroform	Nickel
Formaldehyde	Perchloroethylene
Inorganic arsenic	Trichloroethylene
Methylene chloride	Vinyl chloride

Substances not yet under review:

Acetaldehyde	Hexachlorobenzene
Acrylonitrile	Inorganic lead
Beryllium	Mercury
1,3-Butadiene	N-Nitrosomorpholine
Coke oven emissions	PAHs
Di(2-ethyl hexyl)phthalate	PCBs
Dialkyl-nitrosamines	Propylene oxide
p-Dichlorobenzene	Radionuclides
1,4-Dioxane	Styrene
Environmental tobacco smoke	2,4,6-Trichlorophenol

Compounds for which health information is limited, or not yet sufficient to support review

Acrolein	Manganese
Allyl chloride	Methyl bromide
Benzyl chloride	Methyl chloroform
Chlorobenzene	Nitrobenzene
Chlorophenols	Phenol
Chloroprene	Vinylidene chloride
Cresols	Xylenes
Maleic anhydride	

Source: California Air Resources Board, 1988.

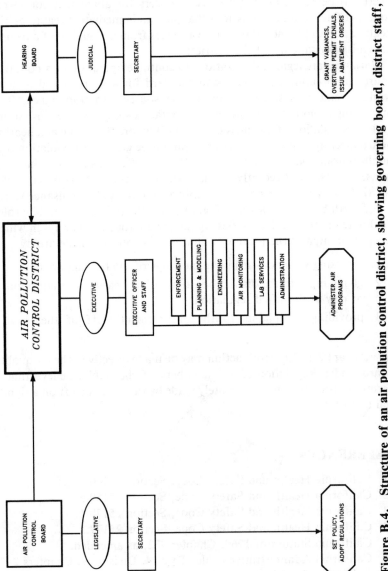

Figure B.4.   Structure of an air pollution control district, showing governing board, district staff, and hearing board (California).

ing actions taken by the district staff (but not the district board). The primary authority of the hearing board is (1) to rule on (and possibly overturn) permit denials; (2) to issue orders for abatement for severe sources of air pollution emissions that pose an immediate threat to public health and safety; and (3) to grant variances from a violation of a technical emission standard of the district.

Variances are legal authorizations to continue operation of a polluting source in violation of an emission standard for a short period of time while a remedy is sought to bring that source into compliance. It is interesting to note that before one can obtain a variance, one must admit guilt to a violation of an emission standard before the public at a hearing board meeting. However, a variance cannot be granted for violation of a public nuisance law[26] such as SCAQMD Rule 402. This has wide implications, because until recently all air toxic regulations not adopted under NESHAPs by local agencies were enforced under public nuisance rules!

A fourth body not shown in Figure B.4 is the public. The public roles are limited in the local air toxics regulatory approach. The ways in which the public impacts the local air pollution regulatory scheme are by:

1. opposing variances at public meetings of the hearing board
2. giving input to the district board at public hearings for adoption of new or modified regulations
3. providing complaints to the enforcement staff of the local agency

Also, an appeal of a permit action may be made directly to the air quality board under Regulation 12[27] if members of the public believe that a permit decision was inappropriately made by the district staff on a significant project.

REFERENCES

1. California Health and Safety Code, Section 25100.
2. California Health and Safety Code, Section 25159.5.
3. California Health and Safety Code, Section 25179.1.
4. California Health and Safety Code, Section 25179.3.
5. California Statues of 1969, Chapter 482, as amended.
6. California Administrative Code, Title 14, Division 7, Chapters 2-5.
7. California Health and Safety Code, Division 20, Chapter 6.6 (November 1986).
8. California Administrative Code, Title 17, Subchapters 1-9.

9. California Vehicle Code, Chapter 382 et seq.
10. California Administrative Code, Title 19.
11. California Health and Safety Code, Section 39650 et seq.
12. "California State Guide to Regulations on Toxics," Office of Administrative Law, Sacramento, CA (1986).
13. California Administrative Code, Title 23, Chapter 3, Subchapter 16.
14. California Health and Safety Code, Division 20, Section 6.7.
15. California Administrative Code, Title 23, Article 4, Chapter 3, Subchapter 16.
16. California Administrative Code, Title 23, Article 4, Chapter 3, Sections 2641–2648.
17. California Administrative Code, Title 23, Section 2631–2634.
18. California Administrative Code, Title 23, Section 2651–2652.
19. California Administrative Code, Title 23, Section 2711–2714.
20. California Administrative Code, Title 22, Section 66694–66696.
21. California Administrative Code, Title 22, Section 66699.
22. California Administrative Code, Title 22, Section 66700.
23. California Administrative Code, Title 22, Section 66702, 66705, 66708.
24. California Health and Safety Code, Section 40002 et seq.
25. California Health and Safety Code, Section 40800 et seq.
26. California Health and Safety Code, Section 42353.
27. Rule 1201, Rules and Regulations of the South Coast Air Quality Management District, El Monte, CA.

GLOSSARY
OF ABBREVIATIONS

AAQ	Ambient air quality
AAQS	Ambient air quality standard
AAS	Atomic absorption spectroscopy
ACGIH	American Conference of Governmental Industrial Hygienists
ADI	Acceptable daily intake (non-inhalation pathway)
AEC	Atomic Energy Commission
APCD	Air Pollution Control District
APHA	American Public Health Agency
API	American Petroleum Institute
AQMD	Air Quality Management District
ASTM	American Society for Testing and Materials
AWWA	American Water Works Association
BACT	Best available control technology (air)
BAT	Best available technology (water)
CAA	Clean Air Act
CAC	California Administrative Code
CAG	Carcinogen Assessment Group (EPA)
CARB	California Air Resources Board
CERCLA	Comprehensive Environmental Response, Compensation and Liability Act
CFR	Code of Federal Regulations
CWA	Clean Water Act
DCE	Dichloroethylene
DES	Diethylstilbestrol
DFA	Department of Food and Agriculture (California)
DHS	Department of Health Services (California)
DNA	Deoxyribonucleic acid
DOT	Department of Transportation
ECD	Electron capture detector

EDB	Ethylene dibromide
EDL	Estimated detection limit
EPA	Environmental Protection Agency
FDA	Food and Drug Administration
FFDCA	Federal Food, Drug and Cosmetic Act
FID	Flame ionization detector
FIFRA	Federal Insecticide, Fungicide and Rodenticide Act
GAC	Granular activated carbon
GC	Gas chromatograph (analyzer)
GC/MS	Gas chromatograph/mass spectrometer (analyzer)
HMTA	Hazardous Materials Transportation Act
HSWA	Hazardous and Solid Waste Act (1984 Amendments to RCRA)
IARC	International Agency for Research on Cancer
ICP	Inductively coupled plasma (analyzer)
LC_{50}	Lethal concentration (inhalation) for 50 percent of the subject population
LD_{50}	Lethal dose (ingestion or dermal exposure) for 50 percent of the subject population
LTD QTY	Limited quantity
MCL	Maximum contaminant level (water)
MCLG	Maximum contaminant level goal (water)
MDL	Method detection limit
MEK	Methyl ethyl ketone
MSDS	Material Safety Data Sheet (OSHA)
NA ID#	North American (continental) identification number
NCP	National Contingency Plan
NESHAP	National Emission Standards for Hazardous Air Pollutants (EPA)
NIOSH	National Institute for Occupational Safety and Health
NOAEL	No Observed Adverse Effect Level
NOEL	No Observed Effect Level
NOIBN	Not otherwise indexed by name
NOS	Not otherwise specified
NPDES	National Pollutant Discharge Elimination System (water)
NPL	National Priorities List
NSPS	New Source Performance Standards (air)
O&M	Operating and maintenance
OSHA	Occupational Safety and Health Administration
OSWER	Office of Solid Waste and Emergency Response (EPA)
PAC	Powdered activated carbon

PAH	Polynuclear aromatic hydrocarbons (also PNAs)
PCB	Polychlorinated biphenyl
PCE	Perchloroethylene
PEL	Permissible Exposure Level
PIC	Product of incomplete combustion
PID	Photoionization detector
POHC	Principal organic hazardous constituent
POTW	Publicly owned treatment works
PPB	Part per billion (concentration)
PPM	Part per million (concentration)
PRP	Potentially responsible party
PVC	Polyvinyl chloride (plastic)
RBC	Rotating biological contactor
RCRA	Resource Conservation and Recovery Act (1976 Amendments to Solid Waste Disposal Act)
RI/FS	Remedial investigation/feasibility study
RO	Reverse osmosis
ROD	Record of Decision (EPA)
RQ	Reportable Quantity
RWQCB	Regional Water Quality Control Board (California)
SARA	Superfund Amendments and Reauthorization Act (1986)
SCAB	South Coast Air Basin (California)
SCAG	Southern California Association of Governments
SCAQMD	South Coast Air Quality Management District (California)
SCM	Suggested Control Measure
SDWA	Safe Drinking Water Act
SEM	Scanning electron microscope
SIP	State Implementation Plan (air quality)
SMEWW	Standard Methods for the Examination of Water and Wastewater
STEL	Short Term Exposure Level (15 minute average)
STLC	Soluble threshold limit concentration (mg/L, California)
SWDA	Solid Waste Disposal Act
SWRCB	State Water Resources Control Board (California)
SW846	EPA test methods for evaluating solid waste—physical and chemical methods
TCA	Trichloroethane
TCDD	Tetrachlorodibenzodioxin; many isomers
TCE	Trichloroethylene
TLV	Threshold Limit Value (inhalation concentration, as established by ACGIH)

TSCA	Toxic Substances Control Act
TSD	Treatment, storage, or disposal
TSDF	Treatment, storage, or disposal facility
TTLC	Total threshold limit concentration (mg/kg, California)
TWA	Time Weighted Average (eight-hour exposure)
UN ID#	United Nations Identification Number
USGS	United States Geologic Survey
VCM	Vinyl chloride monomer
VOC	Volatile organic compound
WET	Waste extraction test (California)
WOGA	Western Oil and Gas Association

INDEX

203

204   HAZARDOUS MATERIALS MANAGEMENT

Resource Conservation and
Recovery Act (RCRA) 152ff.,
173–174 *See also* regulatory
structure
risk
characterization 29–31
definition 21
evaluation 40–41
individual vs societal 22
perceived 22, 40
true value 27
risk assessment
acceptable levels 40 *See also*
Threshold Limit Values
(TLVs)
additivity 30
California methodology 34–40
CARB screening procedure 39
carcinogenicity 23–24, 35
EPA policy 24, 32
guidance documents 24, 34–35,
39
low dose linearity 26–27
major elements 24
models 26–27, 29, 33–38
noncarcinogens 40
populations at risk 31
process 24–31
scaling factors 27
treatment, storage, and disposal
facilities 31–32
uncertainties 31–34
unit risk 29, 39
vs management 23
risk management 22 *See also* waste
minimization
criteria 23, 40 *See also* Threshold
Limit Values (TLVs)
for generators 41

saccharin 21, 32–33
Safe Drinking Water Act (SDWA)
149–150
saturated zone 58ff.
shipping name 85
short-term exposure limit (STEL)
19
sorption 140

species differences *See*
dose-response
state implementation plan (SIP) 50
Superfund Amendments and
Reauthorization Act (SARA)
165–166 *See also* regulatory
structure
synergism 17–18, 31
systemic poisons 10–15, 17

target organs 10–11
teratogen 18, 24
thalidomide 18
Threshold Limit Values (TLVs) 19,
40
toluene 46–49
toxic effects 17
toxicity index 16
toxicity tests *See* dose-response
Toxic Substances Control Act
(TSCA) 52–53, 151–152,
166–167
transporter standards 159
trichloroethylene 46–47, 194
TSDF 31–32, 89, 158–163, 185

underground storage tanks 55, 65,
172, 186–188

vadose zone 58
vinyl chloride 44, 46–47, 49, 52,
194

waste characterization
analytical methods 110ff.
California approaches 188ff.
detection limits 102–104
instrumentation 110ff.
lab methods 111–113
major elements in 98
matrices 101–102
pH 100–101
reference methods 97–99, 108
representative sampling 104–108
sample loss 105–106
samplers 106–109
sampling plans 97, 104–105